T0135782

Bibliografische Information der Deutschen Nationalbibliothek

Die Deutsche Nationalbibliothek verzeichnet diese Publikation in der Deutschen Nationalbibliografie; detaillierte bibliografische Daten sind im Internet über http://dnb.d-nb.de abrufbar.

ISBN 978-3-8325-2466-1

Logos Verlag Berlin GmbH
Comeniushof, Gubener Str. 47,
10243 Berlin
Tel.: +49 (0)30 42 85 10 90
Fax: +49 (0)30 42 85 10 92
INTERNET: http://www.logos-verlag.de

Inverse Problems with Sparsity Constraints: Convergence Rates and Exact Recovery

von Dennis Trede

Dissertation

zur Erlangung des Grades eines Doktors der Naturwissenschaften
— Dr.rer.nat. —

Vorgelegt im Fachbereich 3 (Mathematik & Informatik)
der Universität Bremen
im Dezember 2009

Datum des Promotionskolloquiums:
9. April 2010

Gutachter:
Prof. Dr. Dirk A. Lorenz (Technische Universität Braunschweig)
Prof. Dr. Peter Maaß (Universität Bremen)

Inverse Problems with Sparsity Constraints:

Convergence Rates and Exact Recovery

Dennis Trede

Abstract

This thesis contributes to the field of inverse problems with sparsity constraints. Since the pioneering work by Daubechies, Defries and De Mol in 2004, methods for solving operator equations with sparsity constraints play a central role in the field of inverse problems. This can be explained by the fact that the solutions of many inverse problems have a sparse structure, in other words, they can be represented using only finitely many elements of a suitable basis or dictionary.

Generally, to stably solve an ill-posed inverse problem one needs additional assumptions on the unknown solution—the so-called source condition. In this thesis, the sparseness stands for the source condition, and with that in mind, stability results for two different approximation methods are deduced, namely, results for the Tikhonov regularization with a sparsity-enforcing penalty and for the orthogonal matching pursuit.

The first part of this thesis is the investigation of convergence rates for the sparsity-enforcing Tikhonov regularization in scales of Banach spaces. Classical results for regularization with a quadratic penalty in Hilbert scales are generalized by formulating the penalty and the source condition in a scale of Banach spaces, namely, the scale of Besov spaces.

The second part continues with the Tikhonov regularization, and it goes beyond the question of convergence rates. Assuming that the unknown solution decomposes into a finite number of basis elements, conditions are given, which ensure the exact recovery of those finitely many elements.

In the third part, with almost the same assumptions, conditions for exact recovery for another approximation method are proved, namely, conditions for the orthogonal matching pursuit. Surprisingly, these conditions coincide with those for the Tikhonov regularization deduced before.

Finally, the forth part shows the practical relevance of the conditions for exact recovery. They are used to obtain a priori computable resolution bounds for two examples of convolution type, namely, an example from mass spectrometry and an example from digital holography of particles.

Acknowledgments

I thank Professor Dirk A. Lorenz and Professor Peter Maaß for their supervision and support throughout my graduate studies. I would have been lost without the frequent meetings and the effective discussions we have had.

I am indebted to Theodore Alexandrov, Loïc Denis, Thomas d'Hénin, and Stefan Schiffler for advice and fruitful exchange of ideas. Moreover, I thank Theodore Alexandrov, Christina Brandt, Bastian Kanning, and Stefan Schiffler for careful reading of the manuscript and for giving helpful hints and suggestions. I want to thank all my colleagues at Zentrum für Technomathematik for the good time and the pleasant working atmosphere.

Finally I want to thank my family. I am deeply grateful to my parents Helga and Volker who have supported me all my life, and to my wife Sarah whose love encourages me.

I acknowledge the financial support from the BMBF project INVERS "Deconvolution with sparsity constraints in optical nanoscopy and mass spectroscopy" under the grant 03MAPAH7.

Dennis Trede
Zentrum für Technomathematik
Universität Bremen

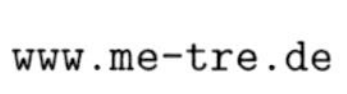

Contents

I Introduction

In the past decades, the field of inverse problems has been a fast growing area of applied mathematics. An inverse problem is a mathematical problem of reconstruction of model parameters from observed data. Inverse problems occur in a wide variety of applications in sciences and engineering, such as medical imaging, remote sensing, nondestructive testing, and astronomy.

Inverse problems often are modeled with an operator equation

$$A(f) = g,$$

where A is a linear or non-linear operator describing the relationship between the data g and model parameters f. The inverse problem consists in the reconstruction of the model parameters f from the knowledge of the forward operator A and the data g. In this thesis we assume A to be a linear operator, and we write $Af = g$. One difficulty of solving the operator equation is, that often only noisy data g^ε are available, and therefore the reconstruction of the model parameters gets unstable. To accomplish that one uses so-called regularization methods.

Since the pioneering work by Daubechies, Defries and De Mol [25], regularization methods for solving operator equations with sparsity constraints play a central role in the field of inverse problems. This can be explained by the fact, that the solutions of many inverse problems have a sparse structure, i.e. f can be represented using only a few elements of a suitable basis or dictionary $\Psi := \{\psi_i\}_i$,

$$f = \textstyle\sum_i u_i \psi_i,$$

where u just exhibits a few non-vanishing coefficients.

The major goal of this thesis is to obtain stability results for regularization methods with sparsity constraints, namely, convergence rates and conditions that ensure exact recovery of the support of u. Two different sparsity-enforcing methods are considered in this thesis:

1. the Tikhonov regularization with sparsity constraints,
2. the orthogonal matching pursuit.

The Tikhonov regularization with sparsity constraints has become popular after the pioneering paper [25]. It is a *variational regularization method* and its idea consists in minimizing the functional

$$T_\alpha(f) := \|Af - g^\varepsilon\|^2 + \alpha \sum_i w_i |\langle f, \psi_i \rangle|^p,$$

with regularization parameter $\alpha > 0$, positive weights $w_i > 0$ and exponent $1 \leq p \leq 2$. Convergence rates for the Tikhonov regularization with a weighted sequence-norm penalty have been deduced in [25,41,65]. There are also convergence rates results for more general penalties [12,49,86,87]. Currently, some first results for the Tikhonov regularization with a nonconvex sequence-norm penalty, namely, for $0 < p < 1$, have been achieved in [9,40,109].

In [78], the author analyzes the classical Tikhonov functional, i.e. the functional with a quadratic penalty, in Hilbert scales. Typical examples of those scales are hilbertian Sobolev spaces. The penalty, a smoothness condition on the operator A, and the source condition are modeled in the scale of Hilbert spaces. A certain a priori parameter rule ensures convergence to a minimum-norm solution with a certain rate. One aim of this thesis is the generalization of these results to scales of Banach spaces, namely, the scale of Besov spaces. Besov spaces coincide with special cases of traditional smoothness function spaces such as Sobolev spaces. However, in contrast to Sobolev spaces the Besov spaces form a scale of Banach spaces in terms of continuous embeddings. In comparison to regularization in Hilbert scales initiated in [78], the relation between these Banach spaces is more complicated.

For the analysis of the Tikhonov regularization in Banach scales we restrict ourselves to reflexive Banach spaces. This limits p to be strictly greater than 1. For the case $p = 1$, in [41] the authors introduce an a priori parameter rule that ensures convergence of the Tikhonov minimizer to a minimum-norm solution with a linear rate. In [36,37,102,103], conditions are given, which ensure the exact recovery of the support of u with $f = \sum_i u_i \psi_i$. The second aim of this thesis is to achieve an a priori parameter rule which ensures exact recovery of the correct support.

The second sparsity-enforcing method considered in this thesis is the orthogonal matching pursuit (OMP). It was first proposed in the signal processing context in [73, 83], as an improvement upon the matching pursuit algorithm [74]. OMP is an *iterative method* that generates a sparse approximate solution by "greedily" minimizing the residual $\|A \cdot -g^{\varepsilon}\|$. The method iteratively selects that element ψ_{i_k} from Ψ, whose corresponding unit-normed image $A\psi_{i_k}/\|A\psi_{i_k}\|$ mostly correlates with the residual in $(k-1)$-st iteration. To stabilize the solution, the iteration has to be stopped early enough.

In comparison with minimization algorithms for Tikhonov functionals with sparsity constraints, the orthogonal matching pursuit is more computationally efficient. However, OMP in general is not a regularization method. There are even undisturbed signals g, for which the algorithm diverges.

In [29, 31, 43, 101], stability results for OMP applied to sparse approximation problems have been presented. The authors provide sufficient conditions, on which OMP can recover the representation of a sparse signal. The concepts from sparse approximation theory are different to those from inverse problems theory, and hence the results from [29, 31, 43, 101] cannot be transfered. The third goal of the thesis is to give exact recovery conditions for OMP, that work for ill-posed inverse problems.

The thesis is organized as follows (see figure 1). In chapter 1, we introduce basic definitions and fundamental results from the theory of inverse problems, which are used in this thesis. We present two classical approaches of regularization methods, namely, *variational regularization methods* and *iterative regularization methods.* Classical regularization methods typically lead to *smooth* approximate solutions. We modify the classical Tikhonov regularization method and the conjugate gradient regularization method in order to achieve a *sparse* approximate solution. These ideas lead to the Tikhonov regularization with a subquadratic penalty and to the greedy iteration called orthogonal matching pursuit, respectively. After this introduction in chapter 1, we pass into the stability analysis of these two sparsity-enforcing methods.

In chapter 2, we investigate regularization properties for the sparsity-enforcing Tikhonov regularization in scales of Besov spaces. We use the Besov scale to model smoothing properties of the operator A, the source condition and the regularization term. With constraints on the Besov-norm penalty term in mind we get a stable approximation and prove a certain convergence rate.

Chapter 3 continues with the sparsity-enforcing Tikhonov regularization. We cite convergence rates results for Tikhonov regularization with a ℓ^1-norm penalty from [41]. The main theoretical results of this chapter go beyond the question of convergence rates. Namely, we deduce conditions which guarantee exact recovery of the unknown support of u.

In chapter 4 we make almost the same assumptions as in chapter 3, and we deduce convergence rates and conditions for exact recovery for the orthogonal matching pursuit. Surprisingly, these conditions coincide with those for the Tikhonov regularization deduced in chapter 3. Moreover, we give an error bound which shows a certain convergence rate.

Finally, in chapter 5 the practical relevance of the conditions for exact recovery is shown. The conditions are used to obtain a priori computable conditions for two examples of convolution type.

In section 5.2 we apply the conditions to an example from mass spectrometry. Here, the data are modeled as sum of Dirac peaks convolved with a Gaussian kernel. To the end of that section we utilize the deduced condition for simulated data of an isotope pattern.

An example from digital holography is considered in section 5.3. The data are modeled as sum of characteristic functions convolved with a Fresnel function. The application of the deduced exact recovery conditions to this examples turns out to be a challenge, because the convolution kernel oscillates. Similar to section 5.2, we apply the theoretical conditions to simulated data, namely, to digital holograms of particles.

The two examples from mass spectrometry and digital holography illustrate that the deduced conditions for exact recovery lead to practically relevant conditions, such that one may check a priori if the experimental setup guarantees exact deconvolution.

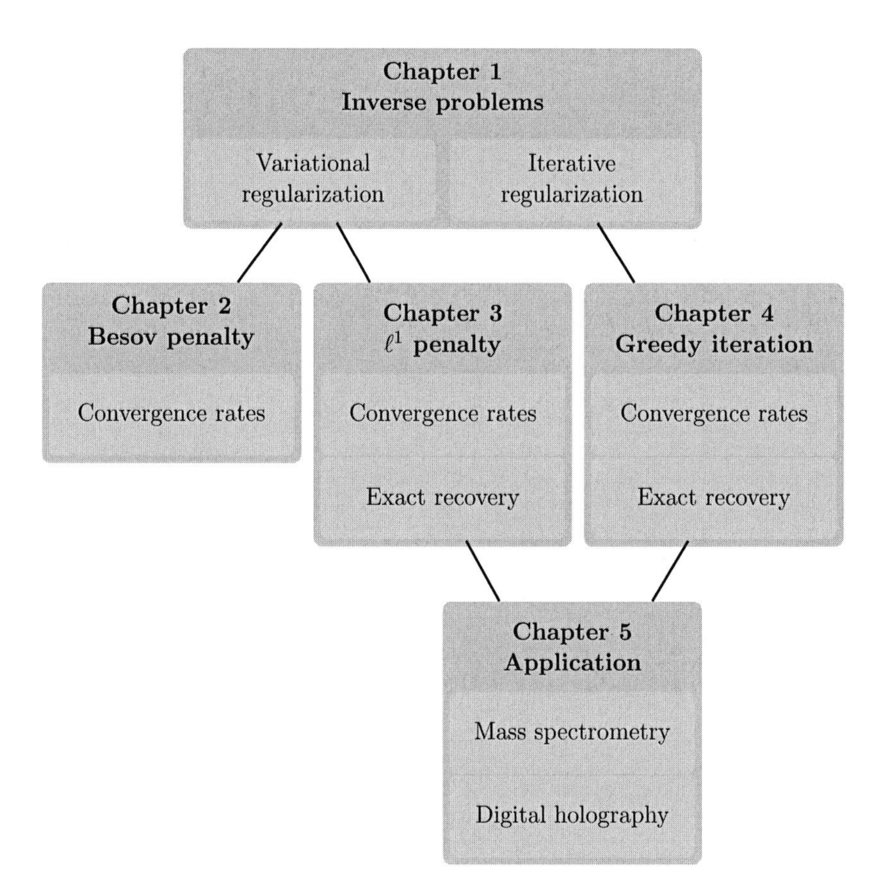

Figure 1: Organization of the thesis.

1 Inverse problems

In this chapter we introduce the basic definitions, the notation and the necessary fundamental results for ill-posed inverse problems, which are used in this thesis. The presentation relies on the textbooks [32, 71, 88] and the introduction of [25].

1.1 Ill-posed inverse problems

In this thesis we consider the linear operator equation of the form

$$Af = g, \tag{1.1}$$

where $A : \mathcal{B}_1 \to \mathcal{H}_2$ is a bounded linear operator between the Banach space $\mathcal{B}_1$ and the Hilbert space $\mathcal{H}_2$. The preimage space $\mathcal{B}_1$ can be a Hilbert space, as well, and in this case it is denoted by $\mathcal{H}_1$.

The *inverse problem* consists in finding an $f \in \mathcal{B}_1$ for given data $g \in \mathcal{H}_2$. If A is injective and surjective, then for any $g \in \mathcal{H}_2$ there exists a unique $f \in \mathcal{B}_1$ with $Af = g$. If additionally A^{-1} is continuous with respect to some topologies in $\mathcal{B}_1$ and $\mathcal{H}_2$, then the solution f depends continuously on the data g. The inverse problem (1.1) is called *well-posed* according to Hadamard's definition, if A satisfies these three properties. If one condition is violated, then the inverse problem (1.1) is called *ill-posed.*

In the following we assume that instead of $g \in \mathrm{rg}(A)$ we are given a noisy observation

$$g^{\varepsilon} = g + \eta$$

with bounded noise η, i.e. $\|g - g^\varepsilon\|_{\mathcal{H}_2} \le \varepsilon$. We try to reconstruct the solution of (1.1) from the knowledge of g^ε. Usually, in ill-posed inverse problems the noisy data g^ε are not in the range of A, and hence the equation (1.1) is not directly solvable. In order to find an estimate of f from observed g^ε, one could possibly minimize the discrepancy

$$\|Af - g^\varepsilon\|^2_{\mathcal{H}_2}. \tag{1.2}$$

This idea leads to the following definition.

Definition 1.1. Let $A : \mathcal{B}_1 \to \mathcal{H}_2$ be a bounded linear operator. The functions that minimize $\|A \cdot - g^\varepsilon\|^2_{\mathcal{H}_2}$ are called *pseudosolutions* of the inverse problem $Af = g^\varepsilon$.

Note that a pseudosolution does not necessarily exist. To see that let $P_{\overline{\mathrm{rg}(A)}}$ and $P_{\mathrm{rg}(A)^\perp} : \mathcal{H}_2 \to \mathcal{H}_2$ denote the orthogonal projection onto $\overline{\mathrm{rg}(A)}$ and $\mathrm{rg}(A)^\perp$, respectively. Then we can rewrite the discrepancy

$$\begin{aligned}\|Af - g^\varepsilon\|^2_{\mathcal{H}_2} &= \|Af - P_{\overline{\mathrm{rg}(A)}}g^\varepsilon\|^2_{\mathcal{H}_2} + \|g^\varepsilon - P_{\overline{\mathrm{rg}(A)}}g^\varepsilon\|^2_{\mathcal{H}_2}\\ &= \|Af - P_{\overline{\mathrm{rg}(A)}}g^\varepsilon\|^2_{\mathcal{H}_2} + \|P_{\mathrm{rg}(A)^\perp}g^\varepsilon\|^2_{\mathcal{H}_2}.\end{aligned}$$

Hence, the minimization of the discrepancy (1.2) is possible if and only if $g^\varepsilon \in \mathrm{rg}(A) \oplus \mathrm{rg}(A)^\perp$. In case of solvability of the minimization problem the set of solutions is given by

$$\mathcal{L}(g^\varepsilon) := \arg\min \|A \cdot - g^\varepsilon\|^2_{\mathcal{H}_2} = \{f \in \mathcal{B}_1 \mid A^*Af = A^*g^\varepsilon\},$$

where A^* denotes the Banach space adjoint operator of A, i.e. the operator

$$A^* : \mathcal{H}_2 \to \mathcal{B}_1^*,$$

which is given by $\langle A^*g, f\rangle_{\mathcal{B}_1^*\times\mathcal{B}_1} = \langle Af, g\rangle_{\mathcal{H}_2}$, where $\langle\cdot,\cdot\rangle_{\mathcal{B}_1^*\times\mathcal{B}_1}$ and $\langle\cdot,\cdot\rangle_{\mathcal{H}_2}$ are dual pairing of $\mathcal{B}_1^* \times \mathcal{B}_1$ and inner product of $\mathcal{H}_2$, respectively.

If A is injective, i.e. the null space of A is trivial, then A^*A is injective. Hence for $g^\varepsilon \in \mathrm{rg}(A) \oplus \mathrm{rg}(A)^\perp$ the set of pseudosolutions $\mathcal{L}(g^\varepsilon)$ is a singleton. If the null space is non-trivial, then one typically chooses an element $f^\dagger \in \mathcal{L}(g^\varepsilon)$ which is minimal in a certain norm.

Definition 1.2. Let $A : \mathcal{B}_1 \to \mathcal{H}_2$ be a bounded linear operator and $g^\varepsilon \in \mathrm{rg}(A) \oplus \mathrm{rg}(A)^\perp$. Moreover, let $\|\cdot\|_{\tilde{\mathcal{B}}}$ be a norm on a subspace $\tilde{\mathcal{B}}$ of

$\mathcal{B}_1$. For $f \in \mathcal{B}_1$ with $f \notin \tilde{\mathcal{B}}$ we define $\|f\|_{\tilde{\mathcal{B}}} := \infty$. Then the functions

$$f^\dagger(g^\varepsilon) \in \arg\min\{\|f\|_{\tilde{\mathcal{B}}} \,|\, f \in \mathcal{L}(g^\varepsilon)\} \tag{1.3}$$

are called *minimum-*$\|\cdot\|_{\tilde{\mathcal{B}}}$ *solutions* of $Af = g^\varepsilon$. If the norm is not specified, then we call it *minimum-norm solution.*

Note that $f^\dagger(g^\varepsilon)$ in general needs not to be unique. If $\ker A = \{0\}$, then $\mathcal{L}(g^\varepsilon)$ is a singleton, and hence, $f^\dagger(g^\varepsilon)$ is unique. For a strictly convex $\|\cdot\|_{\tilde{\mathcal{B}}}$-norm, there is a unique minimum-$\|\cdot\|_{\tilde{\mathcal{B}}}$ solution, since for $g^\varepsilon \in \mathrm{rg}(A) \oplus \mathrm{rg}(A)^\perp$ the set $\mathcal{L}(g^\varepsilon)$ is closed and convex. Hence the minimization problem

$$\min \|f\|_{\tilde{\mathcal{B}}} \quad \text{subject to} \quad A^*Af = A^*g^\varepsilon$$

has a unique solution.

The set of minimum-norm solutions is *not* defined for $g^\varepsilon \in \overline{\mathrm{rg}(A)} \setminus \mathrm{rg}(A)$, since in this case $\mathcal{L}(g^\varepsilon)$ is empty. Even if $g^\varepsilon \in \mathrm{rg}(A) \oplus \mathrm{rg}(A)^\perp$ and if $f^\dagger(g^\varepsilon)$ is unique, it can lead to problems. If the spectrum of A^*A is not bounded below by a strictly positive constant, then the inverse of A^*A is not continuously invertible and hence the mapping

$$g^\varepsilon \mapsto f^\dagger(g^\varepsilon)$$

can be unbounded, too. Thus a small perturbation ε can cause an arbitrarily large error in the related minimum-norm solution $f^\dagger(g^\varepsilon)$. This makes the pseudosolution inappropriate to achieve approximate solutions.

In the following we use the notation $f^\dagger := f^\dagger(g)$ for the minimum-norm solution of the inverse problem $Af = g$ with exact data g. Moreover, in the following we often assume that the equation $Af = g$ attains a solution with certain properties, e.g. smoothness properties. If this knowledge is available we denote the solution with $f^\diamond$. Sometimes $f^\dagger$ and $f^\diamond$ coincide but in general they are different.

1.2 Classical regularization methods

In this section we introduce classical regularization methods for inverse problems. In general, a regularization method approximates an ill-posed problem by a family of neighboring well-posed problems. In the last paragraph we have seen that for noisy data g^ε, the corresponding minimum-norm solution $f^\dagger(g^\varepsilon)$ is not a good approximation for the minimum-norm solution $f^\dagger$ of $Af = g$, even if $f^\dagger(g^\varepsilon)$ exists. For the sake of simplicity in this paragraph the operator A maps from the Hilbert space $\mathcal{H}_1$, i.e. $A : \mathcal{H}_1 \to \mathcal{H}_2$. This setting coincides with the setting of the classical regularization theory.

To make a long story short; a regularization method provides an approximative solution f^ε of a minimum-norm solution $f^\dagger$ of $Af = g$ on basis of the noisy data g^ε such that, on the one hand, f^ε depends continuously on the noisy data g^ε and, on the other hand, f^ε converges to $f^\dagger$ if the noise level ε decreases to zero. This leads to the following definition.

Definition 1.3. Let $f^\dagger$ be a minimum-norm solution of the operator equation $Af = g$ with exact data $g \in \operatorname{rg}(A)$, and let $\alpha_0 \in (0, \infty]$. For every $\alpha \in (0, \alpha_0)$, let

$$\mathcal{R}_\alpha : \mathcal{H}_2 \to \mathcal{H}_1$$

be a continuous operator. Then the family $\{\mathcal{R}_\alpha\}_{0<\alpha<\alpha_0}$ is called a *regularization*, if there exists a *parameter choice rule* $\alpha = \alpha(\varepsilon, g^\varepsilon)$ such that the following holds:

$$\lim_{\varepsilon \to 0} \sup\{\|\mathcal{R}_{\alpha(\varepsilon,g^\varepsilon)}(g^\varepsilon) - f^\dagger\|_{\mathcal{H}_1} \mid g^\varepsilon \in \mathcal{H}_2, \|g - g^\varepsilon\|_{\mathcal{H}_2} \le \varepsilon\} = 0. \quad (1.4)$$

Here, $\alpha : \mathbb{R}^+ \times \mathcal{H}_2 \to (0, \alpha_0)$ is a function such that

$$\lim_{\varepsilon \to 0} \sup\{\alpha(\varepsilon, g^\varepsilon) \mid g^\varepsilon \in \mathcal{H}_2, \|g - g^\varepsilon\|_{\mathcal{H}_2} \le \varepsilon\} = 0. \quad (1.5)$$

The pair $(\mathcal{R}_\alpha, \alpha)$ is called *regularization method*, if equations (1.4) and (1.5) hold.

In summary, the pair $(\mathcal{R}_\alpha, \alpha)$ is called a regularization method if the regularized solution $f^{\alpha,\varepsilon} := \mathcal{R}_{\alpha(\varepsilon,g^\varepsilon)}(g^\varepsilon)$ converges to the minimum-norm solution (in the norm) when the noise level tends to zero. In this thesis we consider convergence in the so-called Bregman distance, as well.

Note that we do not require the regularization operators $\{\mathcal{R}_\alpha\}_{\alpha>0}$ to be a family of *linear* operators. If the operators $\mathcal{R}_\alpha$ are linear, then we call the corresponding method $(\mathcal{R}_\alpha, \alpha)$ a linear regularization method.

By definition 1.3, the parameter rule $\alpha = \alpha(\varepsilon, g^\varepsilon)$ depends on the noise level ε and on the noisy data g^ε. In the following, if α does not depend on g^ε, but only on ε, we call α an *a priori parameter rule* and denote it by $\alpha = \alpha(\varepsilon)$. Otherwise, we call α an *a posteriori parameter rule*.

Another important question for the analysis of regularization methods is, how fast the approximate solution $\mathcal{R}_\alpha(g^\varepsilon)$ approaches the minimum-norm solution $f^\dagger$ as $\varepsilon \to 0$. For a linear operator A which is continuously invertible with $\mathcal{R}_\alpha := A^{-1}$ we achieve linear convergence, so for ill-posed operators we cannot count on better rates than $\mathcal{O}(\varepsilon)$. To prove convergence rates for regularization methods one needs additional knowledge on the regularity of the unknown solution—the so-called *source condition*.

Now the question arises, which operators R_α and parameter choice rules α yield regularization methods? In the following we partially answer it by introducing two examples of classical approaches for regularization methods, namely,

i) *variational regularization methods* represented by the Tikhonov regularization and

ii) *iterative regularization methods* represented by the Landweber iteration and the conjugate gradient regularization method.

1.2.1 Variational regularization methods

In variational regularization, with the regularization parameter $\alpha > 0$, one minimizes a functional

$$T_\alpha : \mathcal{H}_1 \to \mathbb{R},$$

i.e. $\mathcal{R}_\alpha(g^\varepsilon) \in \arg\min_{f \in \mathcal{H}_1} T_\alpha(f)$. One example of variational regularization methods is the classical Tikhonov regularization, which is a popular method for regularization of ill-posed problems.

The idea of the Tikhonov regularization is as follows. If the inverse problem "$Af = g^\varepsilon$" is not directly solvable, instead of minimizing the *discrepancy* $\|A \cdot - g^\varepsilon\|^2_{\mathcal{H}_2}$ we solve the following optimization problem:

$$\min_{f \in \mathcal{H}_1} T_\alpha(f) := \min_{f \in \mathcal{H}_1} \|Af - g^\varepsilon\|^2_{\mathcal{H}_2} + \alpha \|f\|^2_{\mathcal{H}_1}. \tag{1.6}$$

The term $\|\cdot\|^2_{\mathcal{H}_1}$ is called *penalty* and it avoids that the norm of the minimizer gets too large. Hence, in Tikhonov regularization the regularization parameter α represents a trade-off between the energy of f, which is measured in the $\mathcal{H}_1$-norm, and the residual measured in the $\mathcal{H}_2$-norm.

Since, for any $\alpha > 0$, the Tikhonov functional T_α is strictly convex and $\lim_{\|f\| \to \infty} T_\alpha(f) = \infty$, the functional T_α has a unique minimizer.

Let $f^{\alpha,\varepsilon}$ be the unique minimizer of the Tikhonov functional and let $\mathrm{Id} : \mathcal{H}_1 \to \mathcal{H}_1$ denote the identity. Then, from the necessary (and in this case also sufficient) condition

$$\big(\nabla T_\alpha(f)\big)(h) = 0, \quad \text{for all } h \in \mathcal{H}_1,$$

we calculate

$$f^{\alpha,\varepsilon} = (A^*A + \alpha\,\mathrm{Id})^{-1}A^*g^\varepsilon,$$

which can be thought of as a regularized form of the normal equation.

The next question is if minimizing the functional T_α really yields a regularization method. Note that the minimum-$\|\cdot\|_{\mathcal{H}_1}$ solution $f^\dagger$ of $Af = g$ is unique here, since $\mathcal{H}_1$ is a Hilbert space, hence $\|\cdot\|_{\mathcal{H}_1}$ is strictly convex. The next theorem answers this question.

Theorem 1.4 (Stability). *Let $f^{\alpha,\varepsilon}$ be the minimizer of the Tikhonov functional T_α, $g \in \mathrm{rg}(A)$ with $\|g - g^\varepsilon\|_{\mathcal{H}_2} \leq \varepsilon$, and $f^\dagger$ be the minimum-$\|\cdot\|_{\mathcal{H}_1}$ solution of $Af = g$. If the parameter rule $\alpha = \alpha(\varepsilon)$ satisfies*

$$\lim_{\varepsilon\to 0} \alpha(\varepsilon) = 0 \quad \text{and} \quad \lim_{\varepsilon\to 0} \frac{\varepsilon^2}{\alpha(\varepsilon)} = 0, \tag{1.7}$$

then

$$\lim_{\varepsilon\to 0} \|f^{\alpha,\varepsilon} - f^\dagger\|_{\mathcal{H}_1} = 0.$$

Proof. cf. [32, theorem 5.2]. □

Remark 1.5. A simple and commonly used parameter rule that fulfills (1.7), is $\alpha \asymp \varepsilon$. The notation $\asymp$ means that there exist constants $c > 0$ and $C > 0$ such that $c\alpha \leq \varepsilon \leq C\alpha$.

If a certain source condition on the minimum-norm solution $f^\dagger$ is fulfilled, then even a statement on the rate of convergence is possible.

Theorem 1.6 (Convergence rate). *Let $f^{\alpha,\varepsilon}$ be the minimizer of the Tikhonov functional T_α, $g \in \mathrm{rg}(A)$ with $\|g - g^\varepsilon\|_{\mathcal{H}_2} \leq \varepsilon$, and $f^\dagger$ be the minimum-$\|\cdot\|_{\mathcal{H}_1}$ solution of $Af = g$. Moreover, let the following* source condition *be fulfilled, with some $\rho > 0$,*

$$f^\dagger \in \{f \in \mathcal{H}_1 \,|\, \exists w \in \mathcal{H}_1, \|w\|_{\mathcal{H}_1} \leq \rho :\ f = A^*Aw\}. \tag{1.8}$$

If the parameter rule $\alpha = \alpha(\varepsilon)$ satisfies

$$\alpha \asymp \left(\tfrac{\varepsilon}{\rho}\right)^{\frac{2}{3}},$$

then

$$\|f^{\alpha,\varepsilon} - f^{\dagger}\|_{\mathcal{H}_1} = \mathcal{O}\big(\varepsilon^{\frac{2}{3}}\big).$$

Proof. cf. [32, page 120]. □

Remark 1.7. The source condition (1.8) imposes a certain smoothness property on the unknown solution, since typically the operator A^*A has smoothing properties. More general source conditions with $\mu > 0$ and $\rho > 0$ are the so-called *source sets* $\mathcal{X}_{\mu,\rho}$ (cf. [32]),

$$f^{\dagger} \in \mathcal{X}_{\mu,\rho} := \{f \in \mathcal{H}_1 \,|\, f = (A^*A)^{\mu} w,\ \|w\|_{\mathcal{H}_1} \leq \rho\}.$$

The idea of the classical Tikhonov regularization (1.6) can be generalized by minimizing a Tikhonov-type functional

$$T_{\alpha}(f) := \text{distance}(Af, g^{\varepsilon}) + \alpha\, \text{energy}(f),$$

where the functionals

$$\text{distance}(A\cdot, g^{\varepsilon}) : \mathcal{H}_1 \to \mathbb{R} \qquad \text{and} \qquad \text{energy}(\cdot) : \mathcal{H}_1 \to \mathbb{R}$$

measure the discrepancy and the energy, respectively, analogously to the classical Tikhonov regularization.

1.2.2 Iterative regularization methods

In variational regularization methods the regularization parameter α compromises between minimizing the residual and keeping the penalty term small, i.e. enforcing stability. Iterative methods are self-regularizing in the sense that early termination of the iteration process has a regularizing effect. Here, the iteration index constitutes the regularization parameter α and the stopping rule plays the role of the parameter rule.

Iterative regularization methods iteratively minimize the discrepancy $\|Af - g^{\varepsilon}\|_{\mathcal{H}_2}$, starting with an initial guess. For the sake of simplicity in the following we use an initial guess equal to zero.

When applying iterative regularization methods to noisy data g^{ε}, one typically observes a so-called semiconvergence, i.e. at the beginning the reconstruction error decreases but with more iterations it increases. The examples of iterative regularization methods we consider here are the Landweber iteration and the conjugate gradient regularization method.

The Landweber iteration is a simple iterative regularization method and arises from the transformation of the normal equation into an equivalent fixed point equation

$$f = f + \omega A^*(g^\varepsilon - Af).$$

Here, $\omega > 0$ represents a relaxation parameter. Using the initial guess $f^{0,\varepsilon} = 0$, the Landweber iteration computes approximations recursively by

$$f^{k,\varepsilon} := f^{k-1,\varepsilon} + \omega A^*(g^\varepsilon - Af^{k-1,\varepsilon}), \quad k \geq 1.$$

For exact data $g \in \mathrm{rg}(A)$, the sequence $f^{k,0}$ converges to the minimum-$\|\cdot\|_{\mathcal{H}_1}$ solution of $Af = g$. However, if $g^\varepsilon \notin \mathrm{rg}(A) \oplus \mathrm{rg}(A)^\perp$ then the iterates $f^{k,\varepsilon}$ diverge, cf. [88, theorem 5.1.1]. Let $f^\dagger$ denote the minimum-$\|\cdot\|_{\mathcal{H}_1}$ solution of $Af = g$ and split the total error

$$f^\dagger - f^{k,\varepsilon} = f^\dagger - f^{k,0} + f^{k,0} - f^{k,\varepsilon},$$

where $f^{k,\varepsilon}$ and $f^{k,0}$ denote the Landweber iterates with the same relaxation parameter ω for exact data $g \in \mathrm{rg}(A)$ and noisy data $g^\varepsilon \in \mathcal{H}_2$, respectively. Then for $k \to \infty$ the approximation error $f^\dagger - f^{k,0}$ converges to zero and the data error $f^{k,0} - f^{k,\varepsilon}$ diverges if $g^\varepsilon \notin \mathrm{rg}(A) \oplus \mathrm{rg}(A)^\perp$. However, the data error term does not diverge faster than $\sqrt{k\omega}\varepsilon$, cf. [88, theorem 5.1.2]. This effect is called semiconvergence. Thus, the regularizing properties of the Landweber iteration depend on a reliable stopping rule which corresponds to the regularization parameter. The following theorem shows that the iteration should not be stopped too early.

Theorem 1.8. *Let $g \in \mathrm{rg}(A)$ and $g^\varepsilon \in \mathcal{H}_2$ with $\|g - g^\varepsilon\|_{\mathcal{H}_2} \leq \varepsilon$. If $\|Af^{k,\varepsilon} - g^\varepsilon\|_{\mathcal{H}_2} > 2\varepsilon$ and $0 < \omega < \|A\|^{-2}$, then $f^{k+1,\varepsilon}$ is a better approximation than $f^{k,\varepsilon}$, i.e.*

$$\|f^\dagger - f^{k+1,\varepsilon}\|_{\mathcal{H}_1} \leq \|f^\dagger - f^{k,\varepsilon}\|_{\mathcal{H}_1}.$$

Proof. cf. [88, theorem 5.1.4]. □

In other words, the iteration should not be stopped before the following condition is violated:

$$\|Af^{k,\varepsilon} - g^\varepsilon\|_{\mathcal{H}_2} > 2\varepsilon.$$

Moreover, it can be shown that the number of Landweber iterations is finite for the following a posteriori parameter choice rule.

Theorem 1.9 (Stability). *Let $g \in \operatorname{rg}(A)$ and $g^\varepsilon \in \mathcal{H}_2$ with $\|g - g^\varepsilon\|_{\mathcal{H}_2} \leq \varepsilon$. Moreover, let $0 < \omega < \|A\|^{-2}$ and fix $\tau > 1$. Then the discrepancy principle,*

$$\text{find index } k_{\max} \in \mathbb{N}_0 : \quad \|Af^{k_{\max},\varepsilon} - g^\varepsilon\|_{\mathcal{H}_2} \leq \tau\varepsilon < \|Af^{k,\varepsilon} - g^\varepsilon\|_{\mathcal{H}_2}, \tag{1.9}$$

for all $k < k_{\max}$, determines a finite stopping index $k_{\max} = k_{\max}(\varepsilon, g^\varepsilon)$ for the Landweber iteration with $k_{\max}(\varepsilon, g^\varepsilon) = \mathcal{O}(\varepsilon^{-2})$.

Proof. cf. [88, theorem 5.1.5]. □

With a certain source condition on the minimum-norm solution $f^\dagger$ we come to convergence rates.

Theorem 1.10 (Convergence rate). *Let $g \in \operatorname{rg}(A)$, $g^\varepsilon \in \mathcal{H}_2$ with $\|g - g^\varepsilon\|_{\mathcal{H}_2} \leq \varepsilon$, $0 < \omega < \|A\|^{-2}$ and fix $\tau > 1$. Further let $\mu > 0$ and $f^\dagger$ be the minimum-$\|\cdot\|_{\mathcal{H}_1}$ solution of $Af = g$ fulfilling the* source condition

$$f^\dagger \in \operatorname{rg}\left((A^*A)^\mu\right).$$

Then the discrepancy principle (1.9) leads to the convergence rate

$$\|f^{k_{\max},\varepsilon} - f^\dagger\|_{\mathcal{H}_1} = \mathcal{O}\left(\varepsilon^{\frac{2\mu}{2\mu+1}}\right).$$

For the stopping index it holds that $k_{\max}(\varepsilon, g^\varepsilon) = \mathcal{O}(\varepsilon^{-\frac{2}{2\mu+1}})$.

Proof. cf. [32, theorem 6.5]. □

The second iterative regularization method that is considered in this section is the conjugate gradient regularization method. Again, for the sake of simplicity we set $f^{0,\varepsilon} = 0$. Here, the iterates $f^{k,\varepsilon}$ minimize the residual $\|A \cdot - g^\varepsilon\|_{\mathcal{H}_2}$ in the corresponding Krylov subspace $\mathcal{K}_k$, i.e.

$$\|Af^{k,\varepsilon} - g^\varepsilon\|_{\mathcal{H}_2} = \min\{\|Af - g^\varepsilon\|_{\mathcal{H}_2} \mid f \in \mathcal{K}_k\}, \tag{1.10}$$

where the Krylov subspace $\mathcal{K}_k$, with $r^k := Af^{k,\varepsilon} - g^\varepsilon$, is defined as

$$\begin{aligned}\mathcal{K}_k :&= \operatorname{span}\{A^*g^\varepsilon, (A^*A)A^*g^\varepsilon, (A^*A)^2A^*g^\varepsilon, \ldots, (A^*A)^{k-1}A^*g^\varepsilon\} \\ &= \operatorname{span}\{A^*r^0, A^*r^1, \ldots, A^*r^{k-1}\}.\end{aligned} \tag{1.11}$$

The minimization problem (1.10) for $k \geq 1$ can be efficiently solved with the conjugate gradient algorithm.

One advantage of the conjugate gradient method is that the residual of the k-th iterate is always smaller than the corresponding residual of the Landweber iterate, cf. [32, section 7.3]. Thus, if we use the same stopping rule as above for the Landweber iteration,

$$\text{find index } k_{\max} \in \mathbb{N}_0 : \quad \|Af^{k_{\max},\varepsilon} - g^\varepsilon\|_{\mathcal{H}_2} \leq \tau\varepsilon < \|Af^{k,\varepsilon} - g^\varepsilon\|_{\mathcal{H}_2},$$

for all $k < k_{\max}$, then theorem 1.9 guarantees a finite stopping index $k_{\max}$ for the conjugate gradient method, as well. Hence, the conjugate gradient method together with the stopping rule above yields a regularization method.

1.3 Incorporating sparsity constraints

The above introduced variational and iterative regularization methods typically lead to smooth approximate solutions. In [1], the effect of over-smoothing with the classical Tikhonov regularization is discussed in the context of image processing. In some applications, however, one a priori knows that the unknown solution f of $Af = g$ has a sparse representation in a certain basis or dictionary $\Psi = \{\psi_i\}_{i\in\mathbb{Z}}$, i.e. the unknown solution f can be expressed as

$$f = \sum_{i\in\mathbb{Z}} u_i\psi_i,$$

where u exhibits finitely many nonvanishing coefficients.

The knowledge that f has a sparse representation can be used for the reconstruction. In paragraph 1.3.1 we introduce the Tikhonov regularization with sparsity constraints. Here, the classical Tikhonov functional (1.6) is modified so that it enforces a sparse minimizer. In paragraph 1.3.2 we deal with an iterative method that provides sparse approximate solutions. In the following we do not restrict the setting to operators on Hilbert spaces but define A to a be bounded linear operator mapping from a Banach space $\mathcal{B}_1$ to a Hilbert space $\mathcal{H}_2$, i.e. $A : \mathcal{B}_1 \to \mathcal{H}_2$.

1.3.1 Tikhonov regularization with sparsity constraints

The knowledge that f has a sparse representation can be utilized by adding a nonquadratic penalty term to the discrepancy. In [25], the Tikhonov regularization with a weighted ℓ^p-norm penalty, with $1 \leq p \leq$

2, was introduced in the infinite dimensional inverse problems setting to obtain a sparse minimizer. More precisely, given the regularization parameter $\alpha > 0$, a basis $\Psi = \{\psi_i\}_{i\in\mathbb{Z}}$ of $\mathcal{B}_1$ and a sequence of strictly positive weights $\{w_i\}_{i\in\mathbb{Z}}$, in [25] the authors introduce the following functional

$$T_\alpha(f) = \|Af - g^\varepsilon\|^2_{\mathcal{H}_2} + \alpha \sum_{i\in\mathbb{Z}} w_i |\langle f, \psi_i\rangle|^p. \tag{1.12}$$

In contrast to the classical Tikhonov functional with a quadratic penalty, the ℓ^p-penalized functional (1.12), with $p < 2$, promotes sparsity since small coefficients are penalized stronger. To see this, consider the family of ℓ^p-penalized functionals (1.12) for $1 \le p \le 2$. Keeping the regularization parameter α and the weights $\{w_i\}_{i\in\mathbb{Z}}$ fixed, we decrease p from 2 to 1. This gradually increases the penalization of small coefficients (i.e. those with $|\langle f, \psi_i\rangle| < 1$), while large coefficients ($|\langle f, \psi_i\rangle| > 1$) are less penalized at the same time, see figure 2.

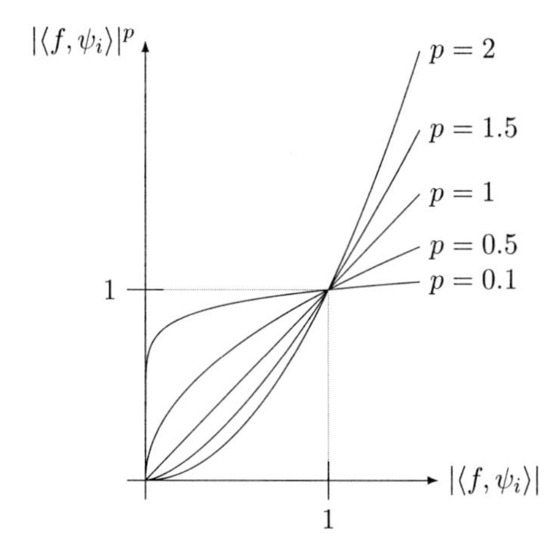

Figure 2: Visualization of ℓ^p penalties with $p \in \{0.1, 0.5, 1, 1.5, 2\}$.

Hence the ℓ^p-penalized Tikhonov functional (1.12), with $p < 2$, promotes sparsity of the expansion of f with respect to the basis elements ψ_i. With this motivation, the penalty $\sum |\langle f, \psi_i\rangle|^p$ with $p < 1$ promotes even more sparsity of the expansion of the minimizer, see figure 2. However, in contrast to $1 \le p \le 2$ the ℓ^p-penalized Tikhonov functionals for $p < 1$ are not convex. For $0 < p < 1$ there exists a minimizer although it need not to be unique. For $p = 0$ the existence of a minimizer is not assured.

In this thesis we restrict ourselves to convex Tikhonov functionals T_α and consider a particular instance of Tikhonov functionals, namely,

$$T_\alpha(f) := \|Af - g^\varepsilon\|^2_{\mathcal{H}_2} + \alpha\|f\|^p_{\mathcal{B}_R}, \tag{1.13}$$

where $\|\cdot\|_{\mathcal{B}}$ denotes the norm of a certain Banach space $\mathcal{B}_R \subset \mathcal{B}_1$. In chapter 2 the Banach space $\mathcal{B}_R$ is defined as Besov space $B^{p,s}_q$, with $1 < p \leq 2$, and in chapter 3 as the sequence space ℓ^1, i.e. $p = 1$. For the Besov space penalty, we utilize the fact that Besov norms can be equivalently expressed using wavelet coefficients. This characterization allows us to represent the Besov penalty as weighted ℓ^p norm.

In section 1.2.1 we have seen that the classical Tikhonov functional with the quadratic penalty yields a regularization method. In chapter 2 and chapter 3 we show similar results for functionals with $p < 2$. For the minimum-norm solution $f^\dagger$ we assume that a certain prior information is available, namely, that $f^\dagger$ has a certain Besov smoothness (chapter 2), or that $f^\dagger$ can be expressed with finitely many nonvanishing basis coefficients (chapter 3). Again, we can describe the Besov smoothness in terms of weighted wavelet coefficients. Exposing one of the source conditions above, we can derive the rate of convergence. For the ℓ^1-penalized functional, we additionally give a condition that ensures recovery of the exact support.

Note that in this thesis we consider a priori parameter rules for Tikhonov functionals. For a posteriori parameter rules for Tikhonov functionals with a subquadratic penalty see e.g. [3, 53].

1.3.2 Greedy iteration

An iterative method that generates a sparse approximate solution from the knowledge of g^ε is concerned in chapter 4. Using the initial guess $f^{0,\varepsilon} = 0$, we minimize the residual $\|A\cdot - g^\varepsilon\|_{\mathcal{H}_2}$ "greedily". For $k \geq 1$, we select iteratively that element ψ_{i_k} from

$$\{\psi_i\}_{i\in\mathbb{Z}},$$

which corresponding unit-normed image $A\psi_{i_k}/\|A\psi_{i_k}\|$ is mostly correlated with the residual in k-th iteration, i.e. with $r^k := Af^{k,\varepsilon} - g^\varepsilon$. To guarantee the well-definedness, we have to ensure that $A\psi_i \neq 0$ for all $i \in \mathbb{Z}$. Similar to the conjugate gradient method (compare equations (1.10) and (1.11)) this iteration can be written by

$$f^{k,\varepsilon} \in \arg\min\{\|Af - g^\varepsilon\|_{\mathcal{H}_2} \mid f \in \mathcal{G}_k\},$$

where the subspace $\mathcal{G}_k$ is defined as

$$\mathcal{G}_k = \operatorname{span}\Big\{ \arg\sup_{\psi_i} |\langle r^0, \tfrac{A\psi_i}{\|A\psi_i\|}\rangle|, \ldots, \arg\sup_{\psi_i} |\langle r^{k-1}, \tfrac{A\psi_i}{\|A\psi_i\|}\rangle| \Big\}.$$

Note that in infinite dimensional Hilbert spaces the supremum

$$\sup_{i\in\mathbb{Z}} \{ |\langle r^k, \tfrac{A\psi_i}{\|A\psi_i\|}\rangle| \}$$

does not have to be realized. Because of that, the method has a variant which only chooses an element that is nearly optimal. We postpone this detail for chapter 4.

This greedy iteration was first proposed in the signal processing context under the name orthogonal matching pursuit (OMP) in [73,83]. In chapter 4 we see that OMP in general does not form a regularization method. However, on certain conditions there is an a posteriori parameter rule that guarantees OMP to be a regularization method. Furthermore, on the same conditions the reconstruction of the exact support is possible.

1.4 Convex functionals

In chapters 2 and 3 we investigate regularization properties of convex Tikhonov functionals. In this paragraph we give some fundamental definitions of convex optimization to provide a basis for this analysis. More facts on convex optimization can be found in the textbook [89].

We consider the following framework. Let $\mathcal{B}_1$ be a Banach space and $R : \mathcal{B}_1 \to [0,\infty]$ be a convex and proper penalty functional. Note that R is allowed to reach infinity and recapitulate that we call R *proper*, if $\operatorname{dom}(R) := \{f \in \mathcal{B}_1 \mid R(f) < \infty\} \neq \emptyset$. Furthermore, remember that the operator $R : \mathcal{B}_1 \to [0,\infty]$ is called *convex*, if for $f_1, f_2 \in \mathcal{B}_1$ and $\gamma \in [0,1]$ the following inequality holds:

$$R(\gamma f_1 + (1-\gamma)f_2) \leq \gamma R(f_1) + (1-\gamma)R(f_2). \tag{1.14}$$

The operator R is called *strictly convex* if inequality (1.14) holds strictly.

A main tool for convex optimization problems is the subgradient. It can be defined for nonconvex functionals, as well. For the sake of convenience we just define it for convex ones.

Definition 1.11. Let $R : \mathcal{B}_1 \to [0, \infty]$ be a convex and proper functional. Then the *subgradient* ∂R at position $f \in \operatorname{dom}(R)$ is a subset of the dual space $\mathcal{B}_1^*$, and it is defined as

$$\partial R(f) := \{f^* \in \mathcal{B}_1^* \,|\, R(\hat{f}) \geq R(f) + \langle f^*, \hat{f} - f\rangle_{\mathcal{B}_1^* \times \mathcal{B}_1},\ \forall \hat{f} \in \mathcal{B}_1\},$$

where $\langle \cdot, \cdot \rangle_{\mathcal{B}_1^* \times \mathcal{B}_1}$ denotes the dual pairing.

If R is Fréchet differentiable, then the subgradient is a singleton and it coincides with its Fréchet derivative, i.e. $\partial R(f) = \{\nabla R(f)\}$.

Example 1.12 (Subgradient of the absolute value)**.** Let $\mathcal{B}_1 = \mathbb{R}$ and $R : \mathbb{R} \to [0, \infty)$, $R(f) := |f|$, be the absolute value function. Then the subgradient of R is

$$\partial R(f) = \begin{cases} \{+1\}, & \text{for } f > 0, \\ [-1, +1], & \text{for } f = 0, \\ \{-1\}, & \text{for } f < 0. \end{cases}$$

The subgradient of the absolute value function is visualized in figure 3.

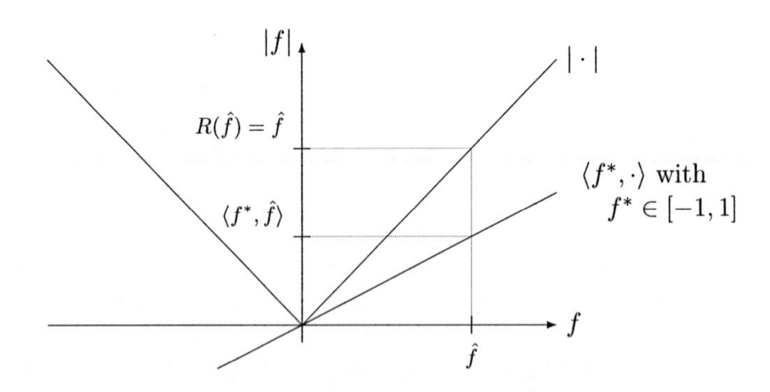

Figure 3: Subgradient of the absolute value function in 0.

In regularization theory in Banach spaces the Bregman distance plays an important role.

Definition 1.13. Let $R : \mathcal{B}_1 \to [0, \infty]$ be a convex and proper functional. Then the *Bregman distance* of R at $f_1 \in \operatorname{dom}(R)$, $\xi_1 \in \partial R(f_1)$ and $f_2 \in \operatorname{dom}(R)$ is defined by

$$D_{\xi_1} : \mathcal{B}_1 \times \mathcal{B}_1 \to [0, \infty],$$
$$D_{\xi_1}(f_2, f_1) := R(f_2) - R(f_1) - \langle \xi_1, f_2 - f_1 \rangle_{\mathcal{B}_1^* \times \mathcal{B}_1}.$$

Analogously, we define the *set-valued Bregman distance* $D_{\partial R(f_1)}(f_2, f_1)$ as the set of all Bregman distances $D_{\xi_1}(f_2, f_1)$ with $\xi_1 \in \partial R(f_1)$.

In figure 4 the Bregman distance is visualized.

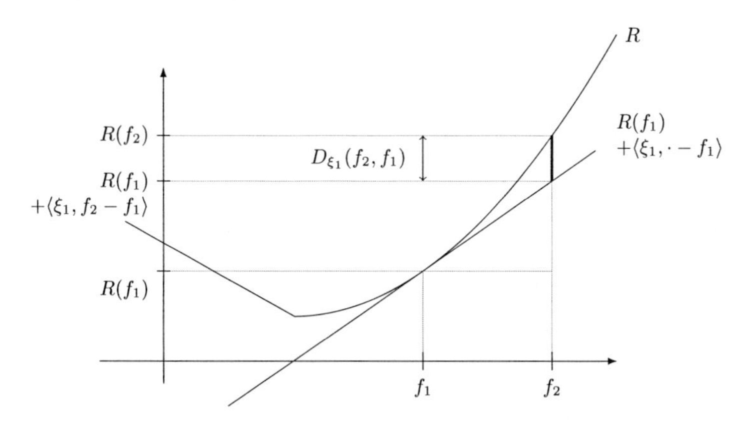

Figure 4: Visualization of the Bregman distance $D_{\xi_1}(f_2, f_1)$.

Remark 1.14. The Bregman distance really is a distance in the sense that

$$D_{\xi_1}(f_2, f_1) \begin{cases} = 0, & \text{if } f_2 = f_1, \\ \geq 0, & \text{else,} \end{cases}$$

since R is convex. However, it does not behave like a metric. Let $f_1, f_2, f_3 \in \operatorname{dom}(R)$, and $\xi_i \in \partial R(f_i)$ for $i = 1, 2, 3$. Then the Bregman distance does not satisfy either the symmetry property, i.e. in general $D_{\xi_1}(f_2, f_1) \neq D_{\xi_2}(f_1, f_2)$, nor the triangle inequality, i.e. in general $D_{\xi_1}(f_3, f_1) \nleq D_{\xi_2}(f_3, f_2) + D_{\xi_1}(f_2, f_1)$. Moreover, $D_{\xi_1}(f_2, f_1) = 0$ does imply $f_2 = f_1$ (only for strictly convex R). More information on Bregman distances can be found in [56].

2 Tikhonov regularization in Besov scales

2.1 Introduction

In this chapter we analyze inverse problems in scales of Banach spaces generalizing classical Hilbert scales. In the classical work [78] by Natterer, the following Tikhonov regularization is considered

$$\min \|Af - g^{\varepsilon}\|^2 + \alpha \|f\|^2_{\mathcal{H}^{s_R}}.$$

Here, the penalty $\|\cdot\|_{\mathcal{H}^{s_R}}$ is the norm in a Hilbert scale $\{\mathcal{H}^s\}_{s\in\mathbb{R}}$. Typical examples for $\mathcal{H}^s$ are hilbertian Sobolev spaces. For the ill-posed operator A a smoothness condition is assumed, which is formulated in the Hilbert scale, namely, that there is a $s_G > 0$ and $M \geq m > 0$ such that for all f it holds that

$$m\|f\|_{\mathcal{H}^{-s_G}} \leq \|Af\| \leq M\|f\|_{\mathcal{H}^{-s_G}}.$$

Moreover, it is assumed that the equation $Af = g$ exhibits a solution $f^{\dagger}$ with $\|f^{\dagger}\|_{\mathcal{H}^{s_S}} \leq \rho$, formulated in the Hilbert space $\mathcal{H}^{s_S}$. With these assumptions in [78] the author comes to an a priori parameter rule and a corresponding error estimates $\|f^{\alpha,\varepsilon} - f^{\dagger}\|_{\mathcal{H}^0}$ for the minimizer

$$f^{\alpha,\varepsilon} = \arg\min \|Af - g^{\varepsilon}\|^2 + \alpha\|f\|^2_{\mathcal{H}^{s_R}}.$$

In this chapter we generalize these regularization results by formulating the penalty, the source condition and the smoothness condition on the operator A in a scale of Banach spaces, namely, the scale of Besov spaces.

Inverse problems formulated in Banach spaces have been of recent interest. There are several theoretical results, such as convergence rates for variational regularization methods in a general Banach spaces setting, see e.g. [12, 49, 86, 87], minimization methods for Tikhonov-type functionals, see e.g. [4, 5, 55], and iterative regularization methods for inverse problems formulated in Banach spaces, see e.g. [45, 46, 54, 95].

The interest in Banach spaces is due to the fact that in many situations a Banach space is better suited to model the data under consideration than a Hilbert space. In the context of image processing, for example, the Banach space BV of functions of bounded variation is used to model images with discontinuities along lines [12, 90, 106]. Moreover, in [49] two examples are presented, in which the use of Banach spaces is necessary for a thorough formulation of the problem. Another class of Banach spaces are the Besov spaces $B_q^{p,s}$ which play an important role in inverse problems related to image processing, see e.g. [18, 19, 64].

A commonly used family of Banach spaces is the family of Sobolev spaces $\{W^{p,s}\}_{p\geq 1, s\in\mathbb{R}}$. We are going to use Besov spaces $B_p^{p,s}$, since they coincide with the Sobolev spaces in most cases if the integrability indices p and s coincide. In contrast to the Sobolev spaces, Besov spaces form a scale of Banach spaces in terms of continuous embeddings. Moreover, they come with a characterization in terms of wavelet coefficients, which make them easy to use for our purposes.

The framework is as follows. Consider the equation

$$Af = g, \tag{2.1}$$

where A is a linear continuous operator

$$A : \mathcal{B}_D \to L^2$$

between the Besov space $\mathcal{B}_D := B_{p_D}^{p_D,s_D}$, with $s_D \in \mathbb{R}$ and $p_D > 1$, and the Lebesgue space L^2. In general these function spaces contain functions or distributions defined on the subset $\Omega \subset \mathbb{R}^d$. For the sake of clarity we omit Ω in the following. The Besov spaces $B_p^{p,s}$ are subspaces of the space of tempered distributions $\mathcal{S}'$ and, in contrast to $\mathcal{S}'$, they are Banach spaces for $p \geq 1$ [91]. Different from classical approaches we use the domain $\mathcal{B}_D$—often a superset of L^2—and not L^2 itself. That is of interest in some applications, e.g. in mass spectrometry where the data consist of delta peaks which are not elements of L^2, see e.g. [57]. For $p \neq \infty$, delta peaks are elements of any Besov space $B_p^{p,s}$ satisfying

$$s < \tfrac{d}{p} - d,$$

where d is the space dimension, see e.g. [91].

If we assume that only noisy data g^ε with noise level $\|g - g^\varepsilon\|_{L^2} \leq \varepsilon$ are available, then the solution of (2.1) could be unstable and has to be stabilized by regularization methods. We use regularization with a Besov constraint, i.e. we regularize by minimizing a not necessarily quadratic Tikhonov functional $T_\alpha : \mathcal{B}_D \to [0, \infty]$ defined by

$$T_\alpha(f) := \|Af - g^\varepsilon\|_{L^2}^2 + \alpha \|f\|_{\mathcal{B}_R}^{p_R}, \tag{2.2}$$

where $\mathcal{B}_R := B_{p_R}^{p_R, s_R} \subset \mathcal{B}_D$ is a Besov space, not necessarily equal to $\mathcal{B}_D$.

In the following we investigate regularization properties and convergence rates of the regularization method consisting of the minimization of (2.2), i.e. $f^{\alpha,\varepsilon} \in \arg\min T_\alpha(f)$. The proceeding is as follows.

i) In section 2.2 we shortly introduce Besov spaces and cite some of their basic properties.

ii) In section 2.3 we reproduce and apply convergence rates results for Banach spaces from [12,49]. With constraints on p_R and s_R in mind and the parameter rule $\alpha \asymp \varepsilon$ we get a stable approximation, i.e.

$$\|f^{\alpha,\varepsilon} - f^\dagger\|_{\mathcal{B}_R} \to 0, \quad \varepsilon \to 0,$$

and a convergence rate in the Sobolev space H^σ

$$\|f^{\alpha,\varepsilon} - f^\dagger\|_{H^\sigma} = \mathcal{O}(\varepsilon^{1/2}),$$

with σ depending on s_R and p_R. These results restrict the choice of possible regularization spaces $\mathcal{B}_R$. Using Besov space embeddings in section 2.4, we get a generalization of the first result. We find a convergence rate—also formulated in a Sobolev space—which holds for a larger set of Besov space penalties $\|\cdot\|_{\mathcal{B}_R}^{p_R}$.

iii) The convergence result gets stronger as σ increases, since for $\theta > 0$ it holds that $H^{\sigma+\theta} \subset H^\sigma$. Since σ depends on s_R and p_R, we address the question how to choose $\mathcal{B}_R$ in a way, such that σ is maximal. We find the regularization penalty $\|\cdot\|_{\mathcal{B}_R}^{p_R}$, which gives the best estimate with respect to σ.

iv) In section 2.5 we apply these results to some operators defined in Sobolev and Besov spaces to demonstrate the differences.

v) Finally, in section 2.6 we give a conclusion and future prospects on Tikhonov regularization in a scale of Banach spaces. Furthermore, we relate the Besov scale results to regularization in Hilbert scales.

The results of this chapter have been published in [67,69].

2.2 Besov spaces and notation

Besov spaces $B_q^{p,s}$ are subspaces of the space of tempered distributions $\mathcal{S}'$. They coincide with special cases of traditional smoothness function spaces, such as Hölder and Sobolev spaces. The following proposition gives a connection of Besov spaces $B_p^{p,s}$ and Sobolev spaces $W^{p,s}$.

Proposition 2.1 (Connection of Besov spaces and Sobolev spaces [91, section 2.1.2]). *The following function spaces have equivalent norms.*

i) $B_p^{p,s} = W^{p,s}$ *for* $s \notin \mathbb{Z}$ *and* $p \geq 1$.

ii) $B_2^{2,s} = H^s := W^{2,s}$ *for all* $s \in \mathbb{R}$.

This clarifies the characterization that the Besov space $B_q^{p,s}$ contains functions having s derivatives in L^p norm. The second integrability index q declares a finer nuance of smoothness. In the following we omit the second integrability index q of the Besov spaces, which is always equal to the corresponding first one p, i.e. $B^{p,s} := B_p^{p,s}$.

In addition, in the following we restrict ourselves to integrability indices $p > 1$. The constraint $p \geq 1$ ensures that $B^{p,s}$ is a Banach space, and if $p > 1$ holds strictly, then it is guaranteed that $B^{p,s}$ is a reflexive Banach space, see proposition 2.5 vide infra.

There are a several ways of defining Besov spaces. Most commonly they are defined via the modulus of smoothness, a way to model differential properties. For a detailed introduction of Besov spaces via moduli of smoothness in conjunction with other smoothness spaces see e.g. [28,100].

Another way of defining Besov spaces is based on wavelet coefficients.

Proposition 2.2 (Equivalent norm in terms of wavelet coefficients [21, 35]). *For all* $s \in \mathbb{R}$, $p > 0$, *there exists an unconditional wavelet basis* $\{\psi_\lambda\}_{\lambda \in \Lambda}$ *such that*

$$\|f\|_{B^{p,s}}^p \asymp \sum_{\lambda \in \Lambda} 2^{p(s+d(\frac{1}{2}-\frac{1}{p}))\,|\lambda|} |u_\lambda|^p, \tag{2.3}$$

where $u_\lambda := \langle f, \psi_\lambda \rangle$ *are the wavelet coefficients of* f.

Recall that the notation $A \asymp B$ means that there exist constants $c > 0$ and $C > 0$ such that $cA \leq B \leq CA$. We use this equivalent norm throughout this chapter.

We shortly remark on bases in Banach spaces. A *Schauder basis* in a Banach space $\mathcal{B}$ is a family of elements $\{\psi_i\}_{i\in\mathbb{Z}}$ such that any $f \in \mathcal{B}$ has a unique expansion $\sum_{i\in\mathbb{Z}} u_i\psi_i$ that converges in $\mathcal{B}$. The basis is called *unconditional*, if the series as well converges, if the coefficients u_i are replaced by v_i with $|v_i| \leq |u_i|$ for $i \in \mathbb{Z}$. This means that one can characterize $\mathcal{B}$ from the size properties of the coefficients in the basis $\{\psi_i\}_{i\in\mathbb{Z}}$. There are classical Banach spaces, in particular $L^1([0,1])$ and $C^0([0,1])$, which have no unconditional basis. For more details on bases in Banach spaces see e.g. [21, 76, 96].

In the following we investigate in regularization properties in scales of function spaces in terms of continuous embeddings. Therefore, for function spaces A_1 and A_2 we have to distinguish between the set-theoretical embedding $A_1 \subset A_2$ and the continuous embedding $A_1 \hookrightarrow A_2$. The continuous embedding $A_1 \hookrightarrow A_2$ means that there exists a constant c such that

$$\|a\|_{A_2} \leq c\,\|a\|_{A_1}, \quad \text{for all } a \in A_1.$$

The following embedding result is an important ingredient in the analysis of the Tikhonov regularization with a Besov penalty.

Proposition 2.3 (Besov scales [91, section 2.2.3]). *Let B^{p_1,s_1} and B^{p_2,s_2} be Besov spaces. If*

$$s_1 - \frac{d}{p_1} > s_2 - \frac{d}{p_2} \quad \textit{and} \quad p_1 \leq p_2, \tag{2.4}$$

then $B^{p_1,s_1} \hookrightarrow B^{p_2,s_2}$. The term $s - \frac{d}{p}$ is called differential dimension *of $B^{p,s}$.*

The embedding of Besov spaces is often visualized with the help of the so-called DeVore diagram [28] where one plots the smoothness s against $1/p$, see figure 5.

Remark 2.4. In contrast to Besov spaces $B^{p,s}$, the family of Sobolev spaces $W^{p,s}$ does not form a scale of Banach spaces in terms of continuous embeddings, cf. [21, section 3.2]. The *Sobolev embedding theorem* states $W^{s_1,p_1} \hookrightarrow W^{s_2,p_2}$ if $s_1 - d/p_1 \geq s_2 - d/p_2$, except for the case where $s_1 - d(1/p_1 - 1/p_2)$ is an integer. Hence there are gaps in the scale of continuous embeddings of Sobolev spaces. This is one reason for using Besov spaces in this chapter.

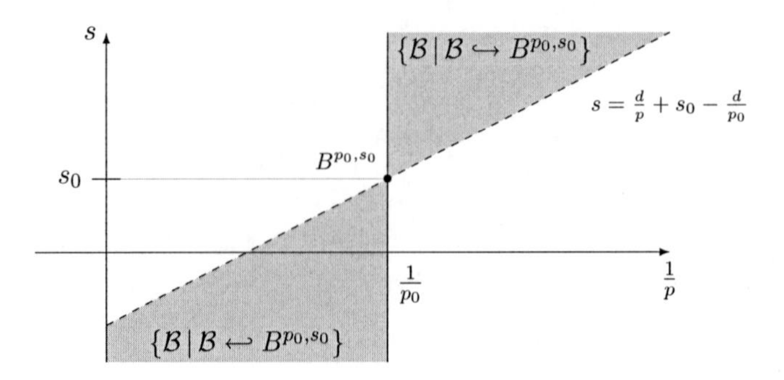

Figure 5: DeVore diagram for the embedding of Besov spaces.

Proposition 2.5 (Duals of Besov spaces [91, section 2.1.5]). *Let* $1 < p < \infty$ *and* $s \in \mathbb{R}$. *Then the dual space of* $B^{p,s}$ *appears as*

$$(B^{p,s})^* = B^{p^*,-s},$$

where p^* *is defined via* $\frac{1}{p} + \frac{1}{p^*} = 1$, *hence* $p^* = \frac{p}{p-1}$.

Assumption 2.6. In this chapter, we use the following Besov spaces. The *domain* of the linear operator A is denoted by

$$\mathcal{B}_D := B^{p_D,s_D},$$

with $s_D \in \mathbb{R}$ and $p_D > 1$. For our *source* $f^\dagger$, we assume that it lives in a Besov space $\mathcal{B}_S$ which is continuously embedded in $\mathcal{B}_D$, i.e.

$$\mathcal{B}_S := B^{p_S,s_S} \hookrightarrow \mathcal{B}_D,$$

hence $s_S - \frac{d}{p_S} > s_D - \frac{d}{p_D}$ and $p_S \leq p_D$. To model smoothing properties of A, we assume that the range of its adjoint A^* is small, namely,

$$\operatorname{rg} A^* = \mathcal{B}_G := B^{p_G,s_G} \hookrightarrow \mathcal{B}_D^* = B^{p_D^*,-s_D}.$$

Consequently, we have $s_G - \frac{d}{p_G} > -s_D - \frac{d}{p_D^*}$ and $\frac{1}{p_G} \geq \frac{1}{p_D^*}$. Later we will see that in fact $\operatorname{rg} A^* \supset \mathcal{B}_G$ is enough, hence one has to choose $\mathcal{B}_G$ as big as possible such that $\mathcal{B}_G \subset \operatorname{rg} A^*$. Another Besov space we have to define is the regularization space

$$\mathcal{B}_R := B^{p_R,s_R} \hookrightarrow \mathcal{B}_D.$$

Since T_α shall be defined on $\mathcal{B}_D$, for $f \in \mathcal{B}_D$ with $f \notin \mathcal{B}_R$ we define $\|f\|_{\mathcal{B}_R} := \infty$. It would be expedient if $\mathcal{B}_S \hookrightarrow \mathcal{B}_R$, since otherwise, if $\mathcal{B}_S \supsetneq \mathcal{B}_R$, then

$$\exists\, f \in \mathcal{B}_S : \quad \|f\|_{\mathcal{B}_R}^{p_R} = \infty,$$

hence $T_\alpha(f) = \infty$. We will see that an appropriate choice of $\mathcal{B}_R$ ensures this property, vide infra in remark 2.13.

Remark 2.7. The smoothness condition on the operator is closely related to Natterer's smoothness condition in Hilbert scales. In this chapter we model smoothness properties with a small range of the adjoint operator, namely, $\operatorname{rg} A^* \supset \mathcal{B}_G$. In [78], a small range of the operator A itself is assumed, namely, $\operatorname{rg} A \subset \mathcal{H}^{s_G}$, see page 23.

2.3 Convergence rates in Besov spaces

2.3.1 Tikhonov regularization in Banach spaces

Tikhonov regularization formulated in Banach spaces has been of recent interest. One important part of the analysis of Tikhonov regularization is the investigation in a priori parameter rules. On the one hand, there are results in a general Banach spaces setting, see e.g. [12,49,86,87]. On the other hand, it has been worked on Tikhonov regularization for special function spaces, such as weighted sequence spaces ℓ^p_w [9,25,40,41,65,109] and the space of functions of bounded variation BV [12].

To come to an a priori parameter rule which yields convergence rates in Besov scales, we use the following results in a general Banach space setting from [49] and [12]. The framework we consider in this subsection is the following. Let $A : \mathcal{B}_1 \to \mathcal{H}_2$ be a linear forward operator between the Banach space $\mathcal{B}_1$ and the Hilbert space $\mathcal{H}_2$. Let the spaces $\mathcal{B}_1$ and $\mathcal{H}_2$ be equipped with topologies $\tau_{\mathcal{B}_1}$ and $\tau_{\mathcal{H}_2}$ which are possibly weaker than the norm topologies. Moreover, let $R : \mathcal{B}_1 \to [0,\infty]$ be a penalty functional. We consider the following Tikhonov functional

$$\|Af - g^\varepsilon\|^2_{\mathcal{H}_2} + \alpha R(f).$$

For A and R let the following assumptions be fulfilled.

(i) $\|\cdot\|_{\mathcal{H}_2}$ is sequentially lower semi-continuous with respect to $\tau_{\mathcal{H}_2}$.

(ii) $A : \mathcal{B}_1 \to \mathcal{H}_2$ is continuous with respect to $\tau_{\mathcal{B}_1}$ and $\tau_{\mathcal{H}_2}$.

(iii) $R : \mathcal{B}_1 \to [0, \infty]$ is proper, convex and $\tau_{\mathcal{B}_1}$ lower semi-continuous.

(iv) $\operatorname{dom}(A) \cap \operatorname{dom}(R) \neq \emptyset$.

(v) For every $\alpha > 0$ and $M > 0$ the following sets are $\tau_{\mathcal{B}_1}$ sequentially compact:

$$\{f \mid \|Af - g^\varepsilon\|_{\mathcal{H}_2}^2 + \alpha R(f) \leq M\}.$$

The following theorem gives a stability result for variational regularization methods in Banach spaces.

Theorem 2.8 (Stability [49, theorem 3.5]). *Let the above assumptions (i)–(v) be fulfilled. Let $\|g - g^\varepsilon\|_{\mathcal{H}_2} \leq \varepsilon$ hold and assume that there exists a unique minimum-R solution $f^\dagger$ of $Af = g$. Then the parameter choice rule $\alpha \asymp \varepsilon$ yields convergence of each minimizer*

$$f^{\alpha,\varepsilon} \in \arg\min \|Af - g^\varepsilon\|_{\mathcal{H}_2}^2 + \alpha R(f)$$

to $f^\dagger$ with respect to $\tau_{\mathcal{B}_1}$.

If a certain prior knowledge is available, then even a statement to the rate of convergence is possible.

Theorem 2.9 (Convergence rate [12, theorem 2]). *Let $\|g - g^\varepsilon\|_{\mathcal{H}_2} \leq \varepsilon$ hold and let $f^\dagger$ be a minimum-R solution of $Af = g$. In addition to the assumptions (i)–(v), assume that the following source condition is satisfied:*

$$\operatorname{rg} A^* \cap \partial R(f^\dagger) \neq \emptyset.$$

Then for the parameter choice rule $\alpha \asymp \varepsilon$ and each minimizer

$$f^{\alpha,\varepsilon} \in \arg\min \|Af - g^\varepsilon\|_{\mathcal{H}_2}^2 + \alpha R(f),$$

there exists an element $d \in D_{\partial R(f^\dagger)}(f^{\alpha,\varepsilon}, f^\dagger)$ such that the following estimate holds:

$$d = \mathcal{O}(\varepsilon).$$

2.3.2 Tikhonov regularization in Besov spaces

The first result we need for the deduction of convergence rates in Besov scales, is a stability result in the regularization space $\mathcal{B}_R$. We prove it by means of theorem 2.8, making use of the equivalent norm (2.3). A similar result with a direct proof can be found in the recent work [84, theorem 3.4].

Theorem 2.10 (Stability in $\mathcal{B}_R$). *Let $\mathcal{B}_R \hookrightarrow \mathcal{B}_D$, $f^\dagger$ be the minimum-$\|\cdot\|_{\mathcal{B}_R}$ solution of $Af = g$, and $\|g - g^\varepsilon\|_{L^2} \le \varepsilon$. Then, for the minimizer $f^{\alpha,\varepsilon}$ of*

$$T_\alpha(f) = \|Af - g^\varepsilon\|_{L^2}^2 + \alpha\|f\|_{\mathcal{B}_R}^{p_R},$$

and the parameter rule $\alpha \asymp \varepsilon$ we get convergence in $\mathcal{B}_R$, i.e.

$$\|f^{\alpha,\varepsilon} - f^\dagger\|_{\mathcal{B}_R} \to 0, \quad \varepsilon \to 0. \tag{2.5}$$

Proof. We equip $\mathcal{B}_D$ and L^2 with the weak topologies and use theorem 2.8. To do so, we need the following to be fulfilled:

(i) The norm $\|\cdot\|_{L^2}$ is weakly lower-semicontinuous in L^2.

(ii) $A : \mathcal{B}_D \to L^2$ is weakly continuous.

(iii) $\|\cdot\|_{\mathcal{B}_R}$ is proper, convex and weakly lower-semicontinuous on $\mathcal{B}_D$.

(iv) $\mathcal{B}_D \cap \mathcal{B}_R \neq \emptyset$.

(v) The sets $\{f \,|\, \|Af - g^\varepsilon\|_{L^2}^2 + \alpha\|f\|_{\mathcal{B}_R}^{p_R} \le M\}$ are weakly sequentially compact in $\mathcal{B}_D$.

The first and the fourth points are obvious. The second point is fulfilled by the assumption that A is linear and continuous. For the fifth point, note that due to the continuous embedding we have $\|\cdot\|_{\mathcal{B}_D} \le c\|\cdot\|_{\mathcal{B}_R}$, and hence the sets are bounded in $\mathcal{B}_D$, which implies weak sequential compactness due to reflexivity of $\mathcal{B}_D$.

For the third point, note that there exists a wavelet basis $\{\psi_\lambda\}$ which is an unconditional basis for both $\mathcal{B}_R$ and $\mathcal{B}_D$. Now, let $u_k \to u$ weakly in $\mathcal{B}_D$ and $u_k \in \mathcal{B}_R$. Since the ψ_λ are also elements of the dual spaces $\mathcal{B}_R^*$ and $\mathcal{B}_D^*$, it holds for all λ that

$$\langle u_k, \psi_\lambda\rangle \to \langle u, \psi_\lambda\rangle, \quad k \to \infty.$$

Hence, a sequence u_k which is bounded in $\mathcal{B}_R$ converges weakly to u in $\mathcal{B}_R$, if it does in the larger space $\mathcal{B}_D$. For that, notice that the duality pairing is the same in both spaces. This shows the weak lower-semicontinuity of $\|\cdot\|_{\mathcal{B}_R}$ on $\mathcal{B}_R$-bounded sets in $\mathcal{B}_D$.

On the condition $p_R > 1$, the $\|\cdot\|_{\mathcal{B}_R}$-norm is strictly convex, so there is a unique minimum-$\|\cdot\|_{\mathcal{B}_R}$ solution $f^\dagger$ of $Af = g$. Now, by theorem 2.8 it follows that $f^{\alpha,\varepsilon} \to f^\dagger$ weakly in $\mathcal{B}_D$ (and by the above considerations also in $\mathcal{B}_R$). Moreover, $\|f^{\alpha,\varepsilon}\|_{\mathcal{B}_R} \to \|f^\dagger\|_{\mathcal{B}_R}$. Since $\mathcal{B}_R$ is uniformly convex, this implies $f^{\alpha,\varepsilon} \to f^\dagger$ strongly in $\mathcal{B}_R$. □

Now we formulate a theorem on the rate of convergence, which follows from theorem 2.9. We assume that a certain knowledge on the minimum-norm solution $f^\dagger$ is available, i.e. a certain source condition is fulfilled. The source condition is formulated in terms of Besov smoothness. This assumption, together with the assumptions on the range of A^*, leads to a regularization term for which a certain convergence rate in a Sobolev norm can be proven.

Theorem 2.11 (Convergence rate in H^σ). *Let* $\operatorname{rg} A^* = \mathcal{B}_G$, $f^\dagger \in \mathcal{B}_S \hookrightarrow \mathcal{B}_D$ *with* $p_S \leq p_G$ *be the minimum-*$\|\cdot\|_{\mathcal{B}_R}$ *solution, and* $\|g - g^\varepsilon\|_{L^2} \leq \varepsilon$. *Then, for the minimizer* $f^{\alpha,\varepsilon}$ *of the Tikhonov functional*

$$T_\alpha(f) = \|Af - g^\varepsilon\|_{L^2}^2 + \alpha \|f\|_{\mathcal{B}_R}^{p_R},$$

with

$$p_R = \frac{p_S + p_G}{p_G}, \quad s_R \leq \frac{p_S s_S - p_G s_G}{p_S + p_G}, \tag{2.6}$$

and the parameter rule $\alpha \asymp \varepsilon$ *we get the convergence rate*

$$\|f^{\alpha,\varepsilon} - f^\dagger\|_{H^\sigma} = \mathcal{O}(\varepsilon^{1/2}), \tag{2.7}$$

where $\sigma := s_R + d\left(\frac{1}{2} - \frac{1}{p_R}\right)$.

Remark 2.12. Notice, that in general the convergence statements in theorem 2.10 and theorem 2.11 correspond to different Besov spaces $\mathcal{B}_R = B^{p_R, s_R}$ and $H^\sigma = B^{2,\sigma}$. The spaces coincide if and only if $p_S = p_G$, hence $p_R = 2$. Otherwise we cannot give any information of inclusions, because the differential dimensions are equal:

$$\sigma - \tfrac{d}{2} = s_R - \tfrac{d}{p_R}.$$

This constitutes the main difference of the results of this chapter and the recently submitted preprint [85], where a convergence rate $\|f^{\alpha,\varepsilon} - f^\dagger\|_{\mathcal{B}_R} = \mathcal{O}(\varepsilon^{1/2})$ is deduced in $\mathcal{B}_R$.

Remark 2.13. The definitions of p_R and s_R in theorem 2.11 imply that $\mathcal{B}_S \hookrightarrow \mathcal{B}_R$. Otherwise the statement would not be meaningful, since if $\mathcal{B}_S \supsetneq \mathcal{B}_R$, then

$$\exists f \in \mathcal{B}_S : \quad \|f\|_{\mathcal{B}_R}^{p_R} = \infty.$$

To see $\mathcal{B}_S \hookrightarrow \mathcal{B}_R$, note that due to $\mathcal{B}_G \hookrightarrow \mathcal{B}_D^*$ the inequality $\frac{1}{p_G} \geq \frac{p_D-1}{p_D}$ holds. Since $\mathcal{B}_S \hookrightarrow \mathcal{B}_D$ we get $\frac{1}{p_S} \geq \frac{1}{p_D}$, and hence

$$p_R = \frac{p_S+p_G}{p_G} = p_S\Big(\frac{1}{p_G} + \frac{1}{p_S}\Big) \geq p_S\Big(\frac{p_D-1}{p_D} + \frac{1}{p_D}\Big) = p_S.$$

To see the inequality for the differential dimension of $\mathcal{B}_R$ and $\mathcal{B}_S$, note that $\mathcal{B}_S \hookrightarrow \mathcal{B}_D \hookrightarrow \mathcal{B}_G^*$, and hence

$$-(s_S + s_G) < d\Big(\frac{p_G-1}{p_G} - \frac{1}{p_S}\Big).$$

This leads to

$$\begin{aligned}-(s_S + s_G)p_G p_S &+ d(p_G + p_S) - d(p_G p_S)\\ &< d\Big(\frac{p_G-1}{p_G} - \frac{1}{p_S}\Big)p_G p_S + d(p_G + p_S) - d(p_G p_S) = 0.\end{aligned}$$

Applying this to the constraints (2.6) for p_R and s_R yields

$$\begin{aligned}s_R - \frac{d}{p_R} &\leq \frac{p_S s_S - p_G s_G}{p_S+p_G} - d\frac{p_G}{p_S+p_G}\\ &= s_S - \frac{d}{p_S} + \frac{-(s_S+s_G)p_G p_S + d(p_G+p_S) - d(p_G p_S)}{(p_S+p_G)p_S} < s_S - \frac{d}{p_S}.\end{aligned}$$

For the proof of theorem 2.11 we need a property of the mapping

$$\|\cdot\|_{\mathcal{B}_R}^{p_R} : \mathcal{B}_S \to [0,\infty).$$

Proposition 2.14. *Let $f \in \mathcal{B}_S$ and let s_R and p_R fulfill (2.6). Then*

$$\partial\Big(\|f\|_{\mathcal{B}_R}^{p_R}\Big) = \Big\{\nabla\|f\|_{\mathcal{B}_R}^{p_R}\Big\} \subset \mathcal{B}_G.$$

Proof. Let $f \in \mathcal{B}_S$ and $\Psi = \{\psi_\lambda\}_{\lambda\in\Lambda}$ be the unconditional wavelet basis of $\mathcal{B}_D$, which is an unconditional basis of $\mathcal{B}_S$, as well, since $\mathcal{B}_S \hookrightarrow \mathcal{B}_R \hookrightarrow \mathcal{B}_D$. Define $u_\lambda := \langle f, \psi_\lambda\rangle$ to be the wavelet coefficients of f. Since $p_S, p_G > 0$, $p_R = 1 + \frac{p_S}{p_G} > 1$, we get

$$\begin{aligned}\partial(\|f\|_{\mathcal{B}_R}^{p_R}) &= \{\nabla\|f\|_{\mathcal{B}_R}^{p_R}\}\\ &= \Big\{p_R \sum_{\lambda\in\Lambda} 2^{p_R(s_R + d(\frac{1}{2} - \frac{1}{p_R}))\,|\lambda|}\,\mathrm{sign}(u_\lambda)\,|u_\lambda|^{p_R-1}\Big\},\end{aligned} \tag{2.8}$$

and hence,

$$\begin{aligned}
&\left\|\nabla\|f\|_{\mathcal{B}_R}^{p_R}\right\|_{\mathcal{B}_G}^{p_G} \\
&= \sum_{\lambda\in\Lambda} 2^{p_G(s_G+d(\frac{1}{2}-\frac{1}{p_G}))\,|\lambda|} \left| p_R\, 2^{p_R(s_R+d(\frac{1}{2}-\frac{1}{p_R}))\,|\lambda|}\, \operatorname{sign}(u_\lambda)\, |u_\lambda|^{p_R-1}\right|^{p_G} \\
&= p_R^{p_G} \sum_{\lambda\in\Lambda} 2^{[p_G(s_G+d(\frac{1}{2}-\frac{1}{p_G}))+p_G p_R(s_R+d(\frac{1}{2}-\frac{1}{p_R}))]\,|\lambda|} |u_\lambda|^{p_G(p_R-1)}.
\end{aligned}$$

Since p_R and s_R satisfy (2.6), we get for the exponent

$$\begin{aligned}
&p_G\big(s_G+d(\tfrac{1}{2}-\tfrac{1}{p_G})\big)+p_G p_R\big(s_R+d(\tfrac{1}{2}-\tfrac{1}{p_R})\big) \\
=&p_G s_G + p_G p_R s_R + d\big(\tfrac{p_G p_R}{2}-\tfrac{p_G}{2}-1\big) \\
=&p_S\big(\tfrac{p_G}{p_S}s_G+\tfrac{p_G}{p_S}s_R+s_R\big)+dp_S\big(\tfrac{1}{2}-\tfrac{1}{p_S}\big) \le p_S s_S + p_S d\big(\tfrac{1}{2}-\tfrac{1}{p_S}\big),
\end{aligned}$$

hence

$$\left\|\nabla\|f\|_{\mathcal{B}_R}^{p_R}\right\|_{\mathcal{B}_G}^{p_G} \le p_R^{p_G} \sum 2^{p_S(s_S+d(\frac{1}{2}-\frac{1}{p_S}))\,|\lambda|} |u_\lambda|^{p_S} \asymp \|f\|_{\mathcal{B}_S}^{p_S} < \infty,$$

since $f \in \mathcal{B}_S$ by assumption. □

Now we are able to prove theorem 2.11.

Proof of theorem 2.11. On the source condition

$$\operatorname{rg} A^* \cap \partial(\|f^\dagger\|_{\mathcal{B}_R}^{p_R}) \neq \emptyset, \tag{2.9}$$

for minimizers $f^{\alpha,\varepsilon}$ of the Tikhonov functional (2.2) and $\alpha \asymp \varepsilon$, theorem 2.9 gives the following estimate for the Bregman distance:

$$D_{\partial\|f^\dagger\|_{\mathcal{B}_R}^{p_R}}(f^{\alpha,\varepsilon}, f^\dagger) = \mathcal{O}(\varepsilon)$$

Here by assumption the range of the adjoint operator A^* is $\mathcal{B}_G$, and hence with proposition 2.14 we get

$$\partial\Big(\|f^\dagger\|_{\mathcal{B}_R}^{p_R}\Big) = \Big\{\nabla\|f^\dagger\|_{\mathcal{B}_R}^{p_R}\Big\} \subset \mathcal{B}_G = \operatorname{rg}(A^*),$$

thus the source condition (2.9) is fulfilled. Note that the assumption $\mathcal{B}_G \subset \operatorname{rg} A^*$ is sufficient here, see assumption 2.6. With $u_\lambda^\dagger$ and $u_\lambda^{\alpha,\varepsilon}$

denoting the wavelet coefficients of $f^\dagger$ and $f^{\alpha,\varepsilon}$, respectively, we get with equation (2.8) that

$$\begin{aligned} D_{\nabla\|f^\dagger\|^{p_R}_{\mathcal{B}_R}}(f^{\alpha,\varepsilon}, f^\dagger) &= \|f^{\alpha,\varepsilon}\|^{p_R}_{\mathcal{B}_R} - \|f^\dagger\|^{p_R}_{\mathcal{B}_R} - \langle \nabla\|f^\dagger\|^{p_R}_{\mathcal{B}_R}, f^{\alpha,\varepsilon} - f^\dagger\rangle \\ &= \sum_{\lambda\in\Lambda} 2^{p_R(s_R+d(\frac{1}{2}-\frac{1}{p_R}))\,|\lambda|}\Big(|u^{\alpha,\varepsilon}_\lambda|^{p_R} - |u^\dagger_\lambda|^{p_R} \\ &\qquad - p_R \operatorname{sign}(u^\dagger_\lambda)\,|u^\dagger_\lambda|^{p_R-1}(u^{\alpha,\varepsilon}_\lambda - u^\dagger_\lambda)\Big) \qquad (2.10) \\ &= \mathcal{O}(\varepsilon). \end{aligned}$$

For $a, b \in \mathbb{R}$, $C > |a|$, $|b-a| < L$, $1 < p \leq 2$ by [7, lemma 4.7] it holds that

$$|b|^p - |a|^p - p \operatorname{sign}(a)|a|^{p-1}(b-a) \geq k(p, C, L)|b-a|^2, \qquad (2.11)$$

where $k(p, C, L)$ is a positive constant which depends on p, C and L. By remark 2.13 it holds that $f^\dagger \in \mathcal{B}_S \hookrightarrow \mathcal{B}_R$, hence

$$\exists C > 0 \quad \forall \lambda \in \Lambda : \quad \Big|2^{(s_R+d(\frac{1}{2}-\frac{1}{p_R}))\,|\lambda|} u^\dagger_\lambda\Big| < C.$$

Furthermore, since $\|f^{\alpha,\varepsilon} - f^\dagger\|_{\mathcal{B}_R} \to 0$ for $\varepsilon \to 0$, we get according to theorem 2.10

$$\exists L > 0 \quad \forall \lambda \in \Lambda : \quad \Big|2^{(s_R+d(\frac{1}{2}-\frac{1}{p_R}))\,|\lambda|}(u^{\alpha,\varepsilon}_\lambda - u^\dagger_\lambda)\Big| < L.$$

Applying this to equation (2.11) with

$$a = 2^{(s_R+d(\frac{1}{2}-\frac{1}{p_R}))\,|\lambda|} u^{\alpha,\varepsilon}_\lambda, \quad b = 2^{(s_R+d(\frac{1}{2}-\frac{1}{p_R}))\,|\lambda|} u^\dagger_\lambda,$$

and $p = p_R \in (1, 2]$ (since $p_G \geq p_S$) we get

$$\begin{aligned} D_{\nabla\|f^\dagger\|^{p_R}_{\mathcal{B}_R}}(f^{\alpha,\varepsilon}, f^\dagger) &\geq \widehat{k} \sum_{\lambda\in\Lambda} 2^{(2(s_R+d(\frac{1}{2}-\frac{1}{p_R})))\,|\lambda|}|u^{\alpha,\varepsilon}_\lambda - u^\dagger_\lambda|^2 \\ &\asymp \|f^{\alpha,\varepsilon} - f^\dagger\|^2_{H^{s_R+d(1/2-1/p_R)}}, \end{aligned}$$

because of the norm equivalence (2.3) and the fact that $H^s = B^{2,s}$ for all s. Finally, this gives

$$\|f^{\alpha,\varepsilon} - f^\dagger\|_{H^\sigma} = \mathcal{O}(\varepsilon^{1/2}),$$

where $\sigma := s_R + d\Big(\frac{1}{2} - \frac{1}{p_R}\Big)$. □

Remark 2.15. To come to an estimate for the Bregman distance (2.10) to above in a certain Besov space norm, i.e.

$$D_{\nabla\|f^\dagger\|_{\mathcal{B}_R}^{p_R}}(f^{\alpha,\varepsilon}, f^\dagger) \geq \widehat{k}\|f^{\alpha,\varepsilon} - f^\dagger\|_{B^{p,s}}^p,$$

we have used equation (2.11). Alternatively one can use the Xu-Roach inequalities for uniformly smooth Banach spaces—a kind of a generalized polarization laws for Banach spaces, cf. [108].

2.4 Regularization in Besov scales

In the setup of theorem 2.11 we assumed that a source condition in terms of Besov smoothness is known, i.e. $f^\dagger \in \mathcal{B}_S$. From that a regularization penalty $\|\cdot\|_{\mathcal{B}_R}^{p_R}$ was derived, which leads to a convergence rate in a certain hilbertian Sobolev space H^σ.

Besov spaces are embedded into each other via the properties (2.4) of proposition 2.3. Considering this, the question arises, which penalties $\|\cdot\|_{\mathcal{B}_R}^{p_R}$ and convergence rates (i.e. which σ) follows from a weakened source condition $f^\dagger \in B^{p,s}$ with $\mathcal{B}_S \hookrightarrow B^{p,s} \hookrightarrow \mathcal{B}_D$. In addition to the embedding properties (2.4) for application of theorem 2.11, one has to ensure $p \leq p_G$. This yields to the following set of possible weaker source conditions

$$\begin{aligned}
f^\dagger \in B^{p,s} \quad \text{such that} \quad & \mathcal{B}_S \hookrightarrow B^{p,s}, \text{ i.e.} & s_S - \tfrac{d}{p_S} &> s - \tfrac{d}{p}, & (2.12)\\
& & \tfrac{1}{p_S} &\geq \tfrac{1}{p}, & (2.13)\\
& B^{p,s} \hookrightarrow \mathcal{B}_D, \text{ i.e.} & s - \tfrac{d}{p} &> s_D - \tfrac{d}{p_D}, & (2.14)\\
& & \tfrac{1}{p} &\geq \tfrac{1}{p_D}, & (2.15)\\
& p_R \in (1,2], \text{ i.e.} & \tfrac{1}{p} &\geq \tfrac{1}{p_G}. & (2.16)
\end{aligned}$$

Figure 6 illustrates the set of weaker source conditions $B^{p,s}$ graphically for $p_D < p_G$.

The direct application of theorem 2.11 to the idea of weakening the source condition with inequalities (2.12)–(2.16) gives the following theorem.

Theorem 2.16 (Convergence rate in Besov scales). *Let* $\operatorname{rg} A^* = \mathcal{B}_G$, $f^\dagger \in \mathcal{B}_S \hookrightarrow \mathcal{B}_D$ *with* $p_S \leq p_G$ *be the minimum-*$\|\cdot\|_{\mathcal{B}_R}$ *solution, and* $\|g - g^\varepsilon\|_{L^2} \leq \varepsilon$. *Further let* $p > 0$ *with* $p_S \leq p \leq \min\{p_D, p_G\}$ *and* $\varsigma > 0$.

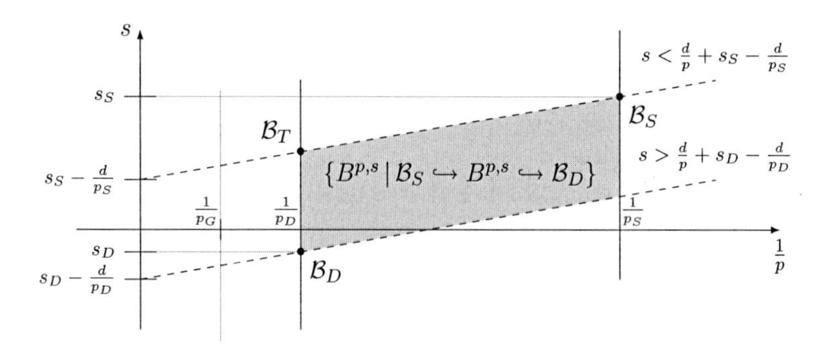

Figure 6: Weaker source conditions $f^\dagger \in B^{p,s}$ with $\mathcal{B}_S \hookrightarrow B^{p,s} \hookrightarrow \mathcal{B}_D$ and $\frac{1}{p_D} > \frac{1}{p_G}$. Corollary 2.17 shows that the highlighted source condition, $f^\dagger \in \mathcal{B}_T \hookleftarrow \mathcal{B}_S$, leads to the best possible convergence rate.

Then, for the minimizer $f^{\alpha,\varepsilon}$ of the Tikhonov functional

$$T_\alpha(f) = \|Af - g^\varepsilon\|_{L^2}^2 + \alpha\|f\|_{\mathcal{B}_R}^{p_R},$$

with

$$p_R = \frac{p + p_G}{p_G}, \quad s_R \leq \frac{p\, s_S - p_G s_G}{p + p_G} - d\Big(\frac{1}{p + p_G}\Big(\frac{p}{p_S} - 1\Big)\Big) - \varsigma, \tag{2.17}$$

and the parameter rule $\alpha \asymp \varepsilon$ we get the convergence rate

$$\|f^{\alpha,\varepsilon} - f^\dagger\|_{H^\sigma} = \mathcal{O}(\varepsilon^{1/2}),$$

where $\sigma := s_R + d\Big(\frac{1}{2} - \frac{1}{p_R}\Big)$.

Proof. From (2.13), (2.15) and (2.16) it follows that theorem 2.11 is applicable for $p > 0$ with $p_S \leq p \leq \min\{p_D, p_G\}$. The choice

$$s \leq s_S - d\big(\tfrac{1}{p_S} - \tfrac{1}{p}\big) - \tilde{\varsigma}$$

with $\tilde{\varsigma} := \frac{p+p_G}{p}\varsigma$ ensures the embedding property (2.12) for the differential dimension. With that the application of theorem 2.11 yields (2.17) and hence the convergence in H^σ. □

The convergence result in theorem 2.16 gets stronger as σ increases. Since σ depends on s_R and p_R, we address the question how to choose $\mathcal{B}_R$ in a way, such that σ is maximal. We try to find that regularization penalty $\|\cdot\|_{\mathcal{B}_R}^{p_R}$ which gives the best estimate with respect to σ (while A and the spaces $\mathcal{B}_S$, $\mathcal{B}_G$ and $\mathcal{B}_D$ are fixed). Since σ depends strictly monotone on s_R, we have to choose s_R as large as possible. Hence we have to solve the following optimization problem which only depends on the exponent p.

$$\left.\begin{aligned} &\max_p \quad \frac{ps_S - p_G s_G}{p+p_G} + d\Big[\frac{1}{2} - \frac{1}{p+p_G}\Big(\frac{p_G}{p_G^*} + \frac{p}{p_S}\Big)\Big] - \varsigma \\ &\text{such that} \quad p_S \le p \le \min\{p_D, p_G\}. \end{aligned}\right\} \tag{2.18}$$

Since $\varsigma > 0$ can be chosen arbitrarily small, we neglect this summand. Hence we have to find the maximum of

$$\begin{aligned} \widehat{\sigma}(p) :&= \tfrac{ps_S - p_G s_G}{p+p_G} - d\big[\tfrac{1}{p+p_G}\big(\tfrac{p_G}{p_G^*} + \tfrac{p}{p_S}\big)\big] \\ &= \tfrac{p}{p+p_G}\big(s_S - \tfrac{d}{p_S}\big) + \tfrac{p_G}{p+p_G}\big(- s_G - \tfrac{d}{p_G^*}\big). \end{aligned}$$

The function $\widehat{\sigma}$ is monotonically increasing in p, since for $p_1 > p_2$ we get

$$\begin{aligned} &\widehat{\sigma}(p_1) - \widehat{\sigma}(p_2) \\ &\quad = \big(\tfrac{p_1}{p_1+p_G} - \tfrac{p_2}{p_2+p_G}\big)\big(s_S - \tfrac{d}{p_S}\big) + \big(\tfrac{p_G}{p_1+p_G} - \tfrac{p_G}{p_2+p_G}\big)\big(- s_G - \tfrac{d}{p_G^*}\big) \\ &\quad = \big(p_G \tfrac{p_1 - p_2}{(p_1+p_G)(p_2+p_G)}\big)\big(s_S - \tfrac{d}{p_S} - \big(- s_G - \tfrac{d}{p_G^*}\big)\big) > 0, \end{aligned}$$

since $\mathcal{B}_S \hookrightarrow \mathcal{B}_D \hookrightarrow \mathcal{B}_G^*$. This proves the following corollary.

Corollary 2.17 (Optimal convergence rate). *Let* $\operatorname{rg} A^* = \mathcal{B}_G$, $f^\dagger \in \mathcal{B}_S \hookrightarrow \mathcal{B}_D$ *with* $p_S \le p_G$ *be the minimum-*$\|\cdot\|_{\mathcal{B}_R}$ *solution,* $\|g - g^\varepsilon\|_{L^2} \le \varepsilon$, *and* $\varsigma > 0$ *small. Then the Tikhonov regularization* T_α *with the parameter rule* $\alpha \asymp \varepsilon$ *and penalty according to (2.17) with*

$$p := \min\{p_D, p_G\}$$

gives the strongest convergence.

i) If $p_G \ge p_D$, *then a penalty with*

$$\begin{aligned} p_R &= \frac{p_D + p_G}{p_G}, \\ s_R &= \frac{p_D s_S - p_G s_G}{p_D + p_G} - d\Big(\frac{1}{p_D + p_G}\Big(\frac{p_D}{p_S} - 1\Big)\Big) - \varsigma, \end{aligned}$$

yields the convergence rate result (2.7) in H^σ *with*

$$\sigma = \frac{p_D s_S - p_G s_G}{p_D + p_G} + d\Big[\frac{1}{2} - \frac{1}{p_D + p_G}\Big(\frac{p_G}{p_G^*} + \frac{p_D}{p_S}\Big)\Big] - \varsigma.$$

ii) If $p_G < p_D$, *then a penalty with*

$$p_R = 2,$$
$$s_R = \frac{1}{2}\Big(s_S - \frac{d}{p_S} - \Big(s_G - \frac{d}{p_G}\Big)\Big) - \varsigma,$$

yields the convergence rate result (2.7) in $\mathcal{B}_R = H^{s_R}$.

Remark 2.18. If $p_S < \min\{p_D, p_G\}$, then the convergence rate in corollary 2.17 is better than in theorem 2.11. Note that the Tikhonov functionals do not coincide.

A curiosity of theorem 2.11, i.e. of the straight forward application of the Banach space regularization results [12,49], is that a more restrictive source condition $\mathcal{B}_T \hookrightarrow \mathcal{B}_S$ does not necessarily enforce a better convergence rate. As the following counterexample shows, sometimes the converse may happen.

Counterexample 2.19. Let $\varsigma > 0$ be sufficiently small and $\tau > 0$. Further let A be an operator with

$$A : H^{-\tau} \to L^2, \quad \operatorname{rg} A^* = H^\tau.$$

1. With the loose source condition $f^\dagger \in \mathcal{B}_S = H^\tau$, theorem 2.11 yields a convergence rate in Lebesgue space L^2. (The choice $\mathcal{B}_S = H^\tau$ leads to $\mathcal{B}_R = L^2$.)

2. If we tighten the condition to $f^\dagger \in B^{1,\tau+\frac{d}{2}+3\varsigma} \hookrightarrow H^\tau$, then we get the regularization space $\mathcal{B}_R = B^{\frac{3}{2},\frac{\tau}{3}+\varsigma+\frac{d}{6}}$ and a convergence rate in Sobolev space $H^{-\frac{\tau}{3}+\varsigma}$, which is larger than in L^2 for small $\varsigma > 0$.

In contrast to that, the usage of Besov space embeddings rewards a tighter source condition with a stronger convergence rate (corollary 2.17). Let $\mathcal{B}_T \hookrightarrow \mathcal{B}_S$, i.e.

$$s_T - \tfrac{d}{p_T} > s_S - \tfrac{d}{p_S} \quad \text{and} \quad p_T \le p_S.$$

Then we get consequently for case *i)* ($p_G \geq p_D$)

$$\sigma(\mathcal{B}_T) - \sigma(\mathcal{B}_S) = \tfrac{p_D}{p_D+p_G}\Big(s_T - \tfrac{d}{p_T} - \big(s_S - \tfrac{d}{p_S}\big)\Big) > 0,$$

and for case *ii)* ($p_G < p_D$)

$$\sigma(\mathcal{B}_T) - \sigma(\mathcal{B}_S) = \tfrac{1}{2}\Big(s_T - \tfrac{d}{p_T} - \big(s_S - \tfrac{d}{p_S}\big)\Big) > 0.$$

2.5 Examples

In the following, we illustrate the convergence rates results with a few examples. With the first one we show that the choice of the parameter p respectively the choice of the source condition $\mathcal{B}_S \hookrightarrow B^{p,s} \hookrightarrow \mathcal{B}_D$ (see equations (2.12)–(2.16)) influences the convergence rate significantly.

Example 2.20 (Smoothing in hilbertian Sobolev scale). Let $d = 1$, $\tau > 1/2$ and consider the operator

$$A : H^{-\tau} \to L^2, \qquad \text{with } \operatorname{rg} A^* = H^{\tau},$$

i.e. we consider smoothing of order τ in the Sobolev scale. Moreover, we assume that the source condition $f^\dagger \in \mathcal{B}_S := B^{1,2\tau} \hookrightarrow H^{-\tau}$ holds, see figure 7.

Theorem 2.16 yields convergence rates for Tikhonov penalties $B^{p_R(p),s_R(p)}$ with $p_S \leq p \leq p_G = p_D$. Since σ is monotone in p we just investigate in the two boundary values here, see solution of the optimization problem (2.18). For $p = p_S$ we get the Tikhonov functional

$$T_\alpha(f) = \|Af - g^\varepsilon\|_{L^2}^2 + \alpha\|f\|_{B^{\frac{3}{2},0}}^{\frac{3}{2}}.$$

With that worst parameter choice respectively that worst source condition theorem 2.16 yields

$$\sigma = \tfrac{2\tau-2\tau}{3} + \tfrac{1}{2} - \tfrac{2}{3} = -\tfrac{1}{6},$$

and hence the convergence rate occurs in a Sobolev space H^σ with negative smoothness.

Next let us check the rate with optimal parameter $p = p_G$, hence $\mathcal{B}_T := H^{2\tau-\frac{1}{2}-\tilde{\varsigma}} \hookrightarrow \mathcal{B}_S$. For the Tikhonov functional

$$T_\alpha(f) = \|Af - g^\varepsilon\|_{L^2}^2 + \alpha\|f\|_{H^{\frac{\tau}{2}-\frac{1}{4}-\varsigma}}^2$$

we get the convergence rate in the Sobolev space H^σ with smoothness

$$\sigma = \tfrac{4\tau-2\tau}{4} + \tfrac{1}{2} - \tfrac{1}{4}(1+2) - \varsigma = \tfrac{\tau}{2} - \tfrac{1}{4} - \varsigma,$$

which is positive for small $\varsigma > 0$, since $\tau > 1/2$. Hence, we get a convergence rate in a Sobolev space with positive smoothness.

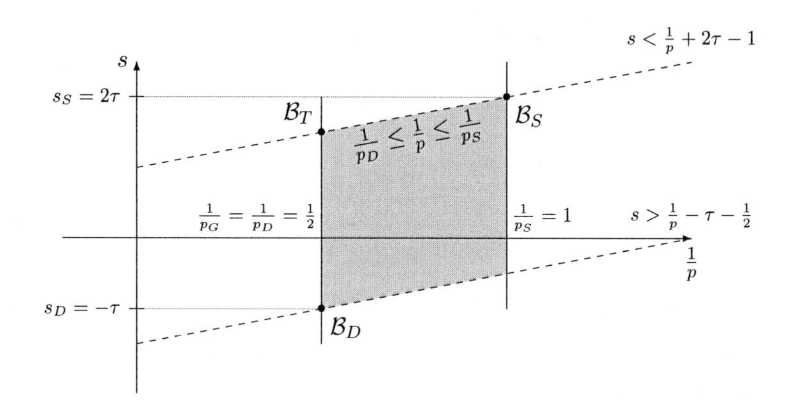

Figure 7: Visualization of the Tikhonov regularization in Besov scales in example 2.20 with source condition $\mathcal{B}_S = B^{1,2\tau}$ and optimal source condition $\mathcal{B}_T = H^{2\tau-\frac{1}{2}-\tilde{\varsigma}}$ with small $\tilde{\varsigma} > 0$.

The first example may lead to the conclusion that a penalty formulated in a Sobolev space gives the best convergence rate. This impression may be intensified, because the optimal source also lives in a Sobolev space, i.e. $p = p_D = 2$. As we see now with the next two examples with operators formulated in Banach scales, this guess is not true. Moreover, the following examples illustrate the difference between the cases $p_S = \min\{p_D, p_G\}$ and $p_S < \min\{p_D, p_G\}$. In the first case, theorem 2.16 yields a convergence rate for only one Tikhonov functional respectively there is no optimization possible (example 2.21). In the second case, we get a set of possible Tikhonov penalties depending on p with $p_S \leq p \leq \min\{p_D, p_G\}$ (example 2.22).

Example 2.21 (Smoothing in the Besov scale with $p_S = \min\{p_D, p_G\}$). Let $d = 1$, $\tau > 0$ and $0 < \theta \leq 1$ small. Consider the operator

$$A : B^{1+\theta,-\tau} \to L^2, \quad \text{with } \operatorname{rg} A^* = B^{\frac{1+\theta}{\theta},\tau} = \left(B^{1+\theta,-\tau}\right)^*,$$

which models smoothing in the scale of Besov spaces. Moreover, let $f^\dagger \in B^{1+\theta,2\tau}$ be the source condition, see figure 8. Notice that $B^{1+\theta,2\tau} \hookrightarrow B^{1+\theta,-\tau}$ and

$$1+\theta \le \tfrac{1+\theta}{\theta}, \quad \text{for } \theta \le 1,$$

and hence we can guarantee that $p_S \le p_G$. Due to $p_S = \min\{p_D, p_G\} = p_D$, it follows from theorem 2.16 that only the Tikhonov functionals with $p = p_S = p_D$ yield a convergence rate in Sobolev space H^σ, i.e. the functionals

$$T_\alpha(f) = \|Af - g^\varepsilon\|_{L^2}^2 + \alpha\|f\|_{\mathcal{B}_R}^{p_R},$$

with $p_R = \frac{p+p^*}{p^*} = p = \theta + 1$ and

$$s_R \le \tfrac{2p\tau - p^*\tau}{p+p^*} = \tfrac{2\theta-1}{\theta+1}\,\tau.$$

The maximal smoothness σ is obtained with the penalty with $s_R = \frac{2\theta-1}{\theta+1}\tau$ and it reads as

$$\sigma = s_R + \tfrac{1}{2} - \tfrac{1}{p_R} = \tfrac{2\theta-1}{\theta+1}\,\tau - \tfrac{1-\theta}{2(1+\theta)}.$$

To put it roughly: For the operator

$$A : B^{1,-\tau} \to L^2 \quad \text{with } \operatorname{rg} A^* = B^{\infty,\tau},$$

the source condition $f^\dagger \in B^{1,2\tau}$ and the Tikhonov functional

$$T_\alpha(f) = \|Af - g^\varepsilon\|_{L^2}^2 + \alpha\|f\|_{B^{1,-\tau}}^1,$$

we a get convergence rate

$$\|f^{\alpha,\varepsilon} - f^\dagger\|_{H^{-\tau-\frac{1}{2}}} = \mathcal{O}(\varepsilon^{1/2}).$$

In the above Besov scale example no optimization of the convergence rate is possible. In the next example there is a set of possible Tikhonov regularizations and hence an optimal one.

Example 2.22 (Smoothing in the Besov scale with $p_S < \min\{p_D, p_G\}$). Let $d = 1$, $\tau > 0$ and $0 < \theta \le 1/2$ small. Consider the operator

$$A : B^{\frac{3}{2},-\tau} \to L^2, \quad \text{with } \operatorname{rg} A^* = B^{3,\tau} = \left(B^{\frac{3}{2},-\tau}\right)^*.$$

Moreover, let $f^\dagger \in \mathcal{B}_S := B^{1+\theta,-\tau+1} \hookrightarrow B^{\frac{3}{2},-\tau}$ be the source condition, see figure 9. Notice that $\mathcal{B}_S \hookrightarrow \mathcal{B}_D$, $p_D < p_G$, and that we can guarantee $p_S \le p_G$. Here theorem 2.16 yields convergence rates for a set

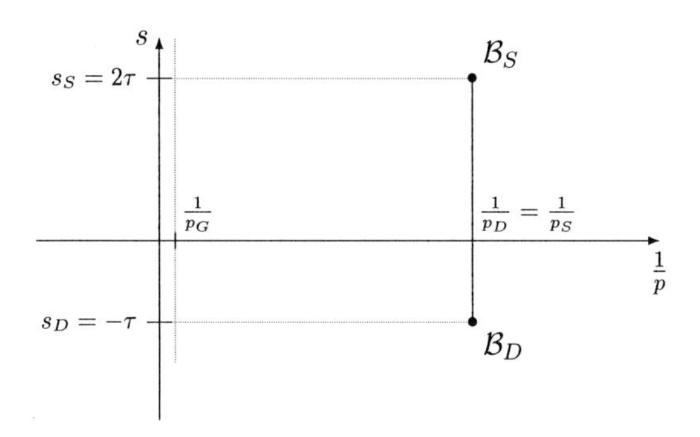

Figure 8: Visualization of the Tikhonov regularization in Besov scales in example 2.21 with source condition $\mathcal{B}_S = B^{1+\theta,\,2\tau}$. Since $p_S = \min\{p_D, p_G\}$ no optimization is possible.

of Tikhonov penalties according to (2.17) with $p_S \le p \le p_D$. We just investigate in the two boundary values again.

With the worst parameter choice, i.e. $p = p_S$, the Tikhonov functional

$$T_\alpha(f) = \|Af - g^\varepsilon\|_{L^2}^2 + \alpha \|f\|_{\mathcal{B}_R}^{p_R},$$

with $p_R = \frac{4+\theta}{3}$ and $s_R = \frac{p_S s_S - p_G s_G}{p_S + p_G} = -\tau + \frac{\theta+1}{\theta+4}$ yields a convergence rate in H^σ with

$$\sigma = -\tau + \tfrac{1}{2} + \tfrac{\theta-2}{\theta+4}.$$

For the optimal parameter $p = p_D$, hence $\mathcal{B}_T := B^{\frac{3}{2},\,-\tau+\frac{2+5\theta}{3+\theta}-\tilde{\varsigma}} \hookrightarrow \mathcal{B}_S$, we get a penalty with $p_R = p_D = \frac{3}{2}$ and

$$s_R = \tfrac{1}{p_G} s_S - \tfrac{1}{p_D} s_G - \left(\tfrac{1}{p_D + p_G}\left(\tfrac{p_D}{p_S} - 1\right)\right) - \varsigma = -\tau + \tfrac{1}{3} + \tfrac{1}{9}\left(\tfrac{2\theta-1}{\theta+1}\right) - \varsigma.$$

Theorem 2.16 yields a convergence rate in H^σ with

$$\sigma = -\tau + \tfrac{1}{6} + \tfrac{1}{9}\left(\tfrac{2\theta-1}{\theta+1}\right) - \varsigma,$$

which is greater than the above σ for $0 < \theta \le 1/2$ and small $\varsigma > 0$.

To put it roughly: Consider the operator

$$A : B^{\frac{3}{2},\,-\tau} \to L^2 \quad \text{with } \operatorname{rg} A^* = B^{3,\,\tau},$$

and the source condition $f^\dagger \in B^{1,\,-\tau+1}$.

For the worst choice $p = p_S$ and the Tikhonov functional

$$T_\alpha(f) = \|Af - g^\varepsilon\|^2_{L^2} + \alpha \|f\|^{\frac{4}{3}}_{B^{\frac{4}{3},-\tau+\frac{1}{4}}},$$

we get a convergence rate

$$\|f^{\alpha,\varepsilon} - f^\dagger\|_{H^{-\tau}} = \mathcal{O}(\varepsilon^{1/2}).$$

The optimal choice $p = p_D$ with the Tikhonov functional

$$T_\alpha(f) = \|Af - g^\varepsilon\|^2_{L^2} + \alpha \|f\|^{\frac{3}{2}}_{B^{\frac{3}{2},-\tau+\frac{2}{9}}},$$

yields a convergence rate

$$\|f^{\alpha,\varepsilon} - f^\dagger\|_{H^{-\tau+\frac{1}{18}}} = \mathcal{O}(\varepsilon^{1/2}).$$

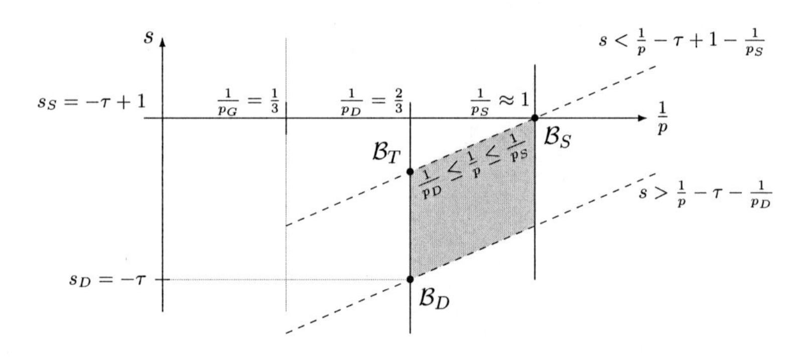

Figure 9: Visualization of the Tikhonov regularization in Besov scales in example 2.22 with small θ. Here the source condition $\mathcal{B}_T$ yields a stronger convergence rate than $\mathcal{B}_S$.

2.6 Conclusion

The aim of this chapter has been to make a first attempt to analyze scales of Banach spaces for Tikhonov regularization. We have used Besov spaces to model the smoothing properties of the operator, the regularization term and the source condition. In comparison to regularization in Hilbert scales initiated in [78], the relation between Besov spaces is more complicated. The convergence rates results have been obtained in the Hilbert scale of Sobolev spaces. Of particular interest is the fact that, on the one hand, tighter source conditions may not lead to stronger convergence rates and, on the other hand, a less tight source condition may result in a stronger result.

Our examples in section 2.5 show only slight improvements in the Sobolev exponents when the Besov-penalty is optimized. It is questionable if the effect can be observed numerically. However, the effect that looser source conditions lead to tighter convergence results is interesting on its own.

Notice that in the Hilbert scale $\{H^s\}_{s\in\mathbb{R}} = \{B^{2,s}\}_{s\in\mathbb{R}}$ no optimization is possible. Assume the operator

$$A : H^{s_D} \to L_2,$$

with smoothing property $\operatorname{rg} A^* = H^{s_G}$ and source condition $f^\dagger \in H^{s_S} \subset H^{s_D}$. In this case, since $p_S = \min\{p_D, p_G\} = 2$, theorem 2.16 just holds for $p = 2$, which yields to convergence in H^σ with $\sigma = s_R$.

So far, we have not used Besov spaces neither to measure the convergence rates nor for the discrepancy term in the Tikhonov functional. Assuming an operator $A : \mathcal{B}_D \to \mathcal{B}_E$ with a certain Besov space $\mathcal{B}_E := B^{p_E,s_E}$ results in the Tikhonov functional

$$T_\alpha(f) := \|Af - g^\varepsilon\|_{\mathcal{B}_E}^{p_E} + \alpha\|f\|_{\mathcal{B}_R}^{p_R}.$$

This setting is of interest and may lead to more general results. To deal with this setting one can use the source conditions from [87] or [49], where variational methods in a Banach-to-Banach space setting are studied. Since the result of this chapter is a first attempt to analyze Tikhonov regularization in Banach scales, we postpone this analysis for future work.

3 Tikhonov regularization with sparsity constraints

3.1 Introduction

In this chapter we consider linear inverse problems with a bounded linear operator $A : \mathcal{H}_1 \to \mathcal{H}_2$ between two separable Hilbert spaces $\mathcal{H}_1$ and $\mathcal{H}_2$,

$$Af = g. \tag{3.1}$$

We are given a noisy observation $g^\varepsilon \in \mathcal{H}_2$ with $\|g - g^\varepsilon\|_{\mathcal{H}_2} \leq \varepsilon$ and try to reconstruct the solution f of $Af = g$ from the knowledge of g^ε.

Moreover, we assume that the operator equation $Af = g$ has a solution $f^\diamond$ that can be expressed sparsely in an orthonormal basis $\Psi := \{\psi_i\}_{i\in\mathbb{Z}}$ of $\mathcal{H}_1$, i.e. $f^\diamond$ decomposes into a finite number of basis elements,

$$f^\diamond = \sum_{i\in\mathbb{Z}} u_i^\diamond \psi_i \quad \text{with} \quad u^\diamond \in \ell^2(\mathbb{Z}, \mathbb{R}),\ \|u^\diamond\|_{\ell^0} < \infty.$$

The uncommon choice of the index set $\mathbb{Z}$ of the basis Ψ accounts for the examples in chapter 5.

The knowledge that $f^\diamond$ can be expressed sparsely can be utilized for the reconstruction by using a *ℓ^1-penalized Tikhonov* regularization, i.e. an approximate solution is given as the minimizer of the functional

$$\tfrac{1}{2}\|Af - g^\varepsilon\|_{\mathcal{H}_2}^2 + \alpha \sum_{i\in\mathbb{Z}} |\langle f, \psi_i\rangle|,$$

with regularization parameter $\alpha > 0$. The Tikhonov regularization with the ℓ^1 penalty was introduced in [25] in the infinite dimensional inverse problems setting to obtain a sparse approximate solution. In contrast to the classical Tikhonov functional with a quadratic penalty,

$$\tfrac{1}{2}\|Af - g^\varepsilon\|_{\mathcal{H}_2}^2 + \alpha \sum_{i\in\mathbb{Z}} |\langle f, \psi_i\rangle|^2,$$

the ℓ^1-penalized functional promotes sparsity since small coefficients are penalized stronger, see section 1.3.1.

For the sake of notational simplification, we introduce the synthesis operator $D : \ell^2 \to \mathcal{H}_1$, which for $u \in \ell^2$ is defined by $Du = \sum u_i\psi_i$. With that and the definition $K := A \circ D : \ell^2 \to \mathcal{H}_2$ we can rewrite the inverse problem (3.1) as $Ku = g$ and the ℓ^1-penalized Tikhonov regularization as

$$T_\alpha(u) := \tfrac{1}{2}\|Ku - g^\varepsilon\|_{\mathcal{H}_2}^2 + \alpha\|u\|_{\ell^1}. \tag{3.2}$$

In the following we frequently use the standard basis of ℓ^2, which is denoted by $\{e_j\}_{j\in\mathbb{Z}}$. Note, that for the synthesis operator D the usage of ℓ^2 is necessary, since any separable Hilbert space is isomorphic to ℓ^2. If $f^\diamond$ was sparse with respect to a dictionary which is not orthogonal, we had to define D for $u \in \ell^1$ to achieve a bounded synthesis operator, compare afterwards in chapter 4.

For minimization of the Tikhonov functional T_α defined in (3.2) there are a lot of more or less efficient minimization algorithms, as e.g. the iterated soft- [8,25] and hard-thresholding [7], the GPSR algorithm [34], and miscellaneous active-set methods, see e.g. [44,61]. For a comparison of some recent algorithms see e.g. [70].

The minimization of the ℓ^1-penalized Tikhonov functional is not the subject of this chapter. We assume that a suitable minimization method is available, and we investigate in error estimates for minimizers $u^{\alpha,\varepsilon} \in \arg\min T_\alpha(u)$, namely, in stability properties, convergence rates and conditions that ensure the exact recovery of the support of $u^\diamond$. The proceeding is as follows.

i) In section 3.2 we give some properties of ℓ^1-penalized Tikhonov minimizers.

ii) In section 3.3 we cite regularization properties and convergence rates from [41], namely, we give an a priori parameter rule that ensures the convergence of $u^{\alpha,\varepsilon} \in \arg\min T_\alpha(u)$ to a minimum-$\|\cdot\|_{\ell^1}$ solution with a certain rate.

iii) Section 3.4 contains the main theoretical results of chapter 3, which go beyond the question of convergence rates. We give a condition which ensures exact recovery of the unknown support of $u^\diamond$. On this condition there exists a regularization parameter α such that for a minimizer $u^{\alpha,\varepsilon} \in \arg\min_{u\in\ell^2} T_\alpha(u)$ it holds that

$$\operatorname{supp}(u^{\alpha,\varepsilon}) = \operatorname{supp}(u^\diamond).$$

Furthermore, we derive this appropriate regularization parameter α in the form of an a priori parameter rule.

iv) In section 3.5 we give a conclusion on exact recovery conditions for Tikhonov regularization with sparsity constraints.

v) Later, in chapter 5, we demonstrate the practicability of the deduced recovery conditions with two examples. In section 5.2 we apply the conditions to an example from mass spectrometry. Here, the data are given as sums of Dirac peaks convolved with a Gaussian kernel. Another example from digital holography is concerned in section 5.3. The data are given as sums of characteristic functions convolved with a Fresnel function. The two examples illustrate that the deduced conditions for exact recovery lead to practically relevant estimates, such that one may check a priori if the experimental setup guarantees exact deconvolution with the regularization method consisting of the minimization of (3.2).

The results of this section have been developed in [66].

3.2 The ℓ^1-penalized Tikhonov functional

Before we start with error estimates, we need some basic properties of the ℓ^1-penalized Tikhonov functional T_α.

Proposition 3.1. *The ℓ^1-penalized Tikhonov functional T_α is convex. It is strictly convex if and only if K is injective.*

Proof. Let v, $w \in \ell^2$ and $\gamma \in [0,1]$. Then, utilizing the linearity of the operator K, we get

$$\begin{aligned}
&T_\alpha(\gamma v + (1-\gamma)w)\\
&= \tfrac{1}{2}\|K(\gamma v + (1-\gamma)w) - g^\varepsilon\|^2_{\mathcal{H}_2} + \alpha\|(\gamma v + (1-\gamma)w)\|_{\ell^1}\\
&= \tfrac{1}{2}\|\gamma(Kv - g^\varepsilon) + (1-\gamma)(Kw - g^\varepsilon)\|^2_{\mathcal{H}_2} + \alpha\textstyle\sum_{i\in\mathbb{Z}} |(\gamma v_i + (1-\gamma)w_i)|.
\end{aligned}$$

The second term can be estimated with the triangle-inequality by

$$\alpha \sum_{i \in \mathbb{Z}} |(\gamma v_i + (1-\gamma) w_i)| \leq \alpha \sum_{i \in \mathbb{Z}} \gamma |v_i| + (1-\gamma)|w_i|.$$

Note, that the absolute value function $|\cdot| : \mathbb{R} \to \mathbb{R}$ is piecewise linear, hence it is not strictly convex. Since the function $(\cdot)^2 : \mathbb{R} \to \mathbb{R}$ is strictly convex, we can estimate the first term as follows,

$$\begin{aligned} &\|\gamma\big(Kv - g^\varepsilon\big) + (1-\gamma)\big(Kw - g^\varepsilon\big)\|^2_{\mathcal{H}_2} \\ &\qquad \leq \gamma \|Kv - g^\varepsilon\|^2_{\mathcal{H}_2} + (1-\gamma)\|Kw - g^\varepsilon\|^2_{\mathcal{H}_2}, \end{aligned}$$

hence the functional T_α is convex.

However, in general T_α is not strictly convex. If K is not injective, then there are v, $w \in \ell^2$ with $Kv = Kw$, hence

$$\|\gamma\big(Kv - g^\varepsilon\big) + (1-\gamma)\big(Kw - g^\varepsilon\big)\|^2_{\mathcal{H}_2} = \|Kv - g^\varepsilon\|^2_{\mathcal{H}_2}.$$

To ensure strict convexity, the inequality

$$\begin{aligned} &\|\gamma\big(Kv - g^\varepsilon\big) + (1-\gamma)\big(Kw - g^\varepsilon\big)\|^2_{\mathcal{H}_2} \\ &\qquad < \gamma \|Kv - g^\varepsilon\|^2_{\mathcal{H}_2} + (1-\gamma)\|Kw - g^\varepsilon\|^2_{\mathcal{H}_2} \end{aligned}$$

has to hold strictly for all v, $w \in \ell^2$. This is the case if and only if K is injective. □

Remark 3.2. In contrast to the considered ℓ^1-penalized functional, the ℓ^p-penalized Tikhonov functionals for $p < 1$ are not convex. For $0 < p < 1$ there exists a minimizer, although it need not to be unique. For $p = 0$ the existence of a minimizer is not assured. In the recent works [9, 40, 109] the authors present stability results for the ℓ^p-penalized Tikhonov regularization with $0 < p < 1$.

Proposition and Definition 3.3 (Optimality condition). *Define the* set-valued sign function

$$\text{Sign} : \ell^2 \to \big\{\{-1\}, [-1,+1], \{+1\}\big\}^{\mathbb{Z}},$$

for $u \in \ell^2$, *by*

$$\big(\text{Sign}(u)\big)_k := \begin{cases} \{-1\}, & u_k < 0, \\ [-1,+1], & u_k = 0, \\ \{+1\}, & u_k > 0. \end{cases} \tag{3.3}$$

Let $u^{\alpha,\varepsilon} \in \ell^2$. Then the following statements are equivalent:

$$\text{(i)} \qquad u^{\alpha,\varepsilon} \in \arg\min_{u \in \ell^2} \tfrac{1}{2}\|Ku - g^\varepsilon\|^2_{\mathcal{H}_2} + \alpha\|u\|_{\ell^1}. \tag{3.4}$$

$$\text{(ii)} \qquad -K^*(Ku^{\alpha,\varepsilon} - g^\varepsilon) \in \alpha\,\mathrm{Sign}(u^{\alpha,\varepsilon}). \tag{3.5}$$

Proof. For a convex functional $R : \ell^2 \to \mathbb{R} \cup \{\infty\}$ it holds that

$$u^* \in \arg\min R \iff 0 \in \partial R(u^*),$$

where $\partial R \subset \ell^2$ denotes the subgradient of R. Since $\|K \cdot -g^\varepsilon\|^2_{\mathcal{H}_2}$ is Fréchet differentiable, for the Tikhonov functional T_α it holds that

$$\begin{aligned} u^{\alpha,\varepsilon} \in \arg\min T_\alpha &\iff 0 \in \{\nabla\|Ku^{\alpha,\varepsilon} - g^\varepsilon\|^2_{\mathcal{H}_2}\} + \alpha\,\partial\|u^{\alpha,\varepsilon}\|_{\ell^1} \\ &\iff -K^*(Ku^{\alpha,\varepsilon} - g^\varepsilon) \in \alpha\,\partial\|u^{\alpha,\varepsilon}\|_{\ell^1}. \end{aligned}$$

To see the equivalence of (3.4) and (3.5) we have to evaluate the subgradient of the ℓ^1 norm. Formally, it can be written as the the set-valued sign function, cf. example 1.12, page 20. However, $\partial\|\cdot\|_{\ell^1}$ is just defined on $\mathrm{dom}(\|\cdot\|_{\ell^1}) = \ell^1$, and not overall ℓ^2. Moreover, in $\ell^1 \setminus \ell^0$ the subgradient is empty. Hence we have to show that condition (3.5) cannot be fulfilled for $u^{\alpha,\varepsilon} \in \ell^2 \setminus \ell^0$.

Note that $\mathrm{rg}(K^*) \subset \ell^2$, hence the left-hand side of condition (3.5) is an element of ℓ^2. However, for $u \in \ell^2 \setminus \ell^0$ it holds that $\mathrm{Sign}(u) \subset \ell^\infty \setminus \ell^2$, since infinitely many coefficients of u do not vanish. Thus, the right-hand side of condition (3.5) is a subset of $\ell^\infty \setminus \ell^2$. With that argument the equivalence of (3.4) and (3.5) follows. □

Proposition 3.4 (Finitely supported Tikhonov minimizer). *Let $u^{\alpha,\varepsilon} \in \arg\min T_\alpha(u)$. Then $u^{\alpha,\varepsilon} \in \ell^0$.*

Proof. According to the proof of proposition and definition 3.3, the optimality condition (3.5) cannot be fulfilled for $u^{\alpha,\varepsilon} \in \ell^2 \setminus \ell^0$. □

In the following the set-valued sign function defined in equation (3.3) is denoted with a capital letter. For the singleton sign function we use the lower case letter and, for $u \in \ell^\infty$, define it by

$$\big(\mathrm{sign}(u)\big)_k := \begin{cases} -1, & u_k < 0, \\ 0, & u_k = 0, \\ 1, & u_k > 0. \end{cases} \tag{3.6}$$

Since the Tikhonov minimizer is finitely supported, we can rewrite the optimality condition (3.5). Let $J \subset \mathbb{Z}$ denote the support of $u^{\alpha,\varepsilon} \in \arg\min T_\alpha$. Then with the scalar sign function, $\operatorname{sign} : \mathbb{R} \to \{-1, 0, 1\}$, the necessary and sufficient condition (3.5) reads as

$$\left.\begin{aligned} -(K^*(Ku^{\alpha,\varepsilon} - g^\varepsilon))_j &= \alpha \operatorname{sign}(u_j^{\alpha,\varepsilon}), && j \in J, && \text{(a)} \\ |K^*(Ku^{\alpha,\varepsilon} - g^\varepsilon)|_j &\leq \alpha, && j \in J^{\complement}. && \text{(b)} \end{aligned}\right\} \quad (3.7)$$

In chapter 2 the Besov penalty with $p_R > 1$ has yielded a *strictly* convex Tikhonov functional, hence a unique minimizer. Here the functional is convex, but in general not strictly convex. Hence the minimizer has not to be unique. In proposition 3.1 we have seen that injectivity of the operator K implies strict convexity of the Tikhonov functional, and hence a unique minimizer is assured. A weaker property of the operator K, which guarantees uniqueness (although the functional is not strictly convex) gives the following definition. The property was introduced in [8], and it is important for the deduction of convergence rates and exact recovery conditions.

Definition 3.5. Let $K : \ell^2 \to \mathcal{H}$ be an operator mapping into a Hilbert space $\mathcal{H}$. Then K has the *finite basis injectivity (FBI)* property, if for all finite subsets $J \subset \mathbb{Z}$ the operator restricted to J is injective, i.e. for all $u, v \in \ell^2$ with $Ku = Kv$ and $u_k = v_k = 0$, for all $k \notin J$, it follows that $u = v$.

In inverse problems with sparsity constraints the FBI property is used for a couple of theoretical results for the ℓ^1-penalized Tikhonov functional, cf. e.g. [22, 41, 42, 44, 52, 65]. The assumption that K is even "fully injective" is more common, but we use the FBI property anyway. Obviously, an injective operator possesses the FBI property, and hence with the FBI property we make a less restrictive assumption. A demonstrative example for an operator which possesses the FBI property but is not fully injective is the synthesis operator which maps wavelet and Gabor coefficients to a signal which is superposition of wavelets and sinusoids.

The FBI property is related to the so-called *restricted isometry property (RIP)* [15] of a matrix, which is a quite common assumption in the theory of compressive sampling [16]. The RIP is defined as follows. Let K be a $m \times n$ matrix and let $N < n$ be an integer. The *restricted isometry constant* of order N is defined as the smallest number $0 < c_N < 1$, such that the following condition holds for all $v \in \mathbb{R}^n$ with at most N non-zero entries:

$$(1 - c_N)\|v\|_{\ell^2}^2 \leq \|Kv\|_{\ell^2}^2 \leq (1 + c_N)\|v\|_{\ell^2}^2.$$

The matrix K has the RIP, if the constants c_N stay reasonably small for a reasonable range of N. Essentially, this property denotes that the matrix is approximately an isometry when restricted to small subspaces. The FBI property, however, is defined for operators acting on the sequence space and only says, that the restriction to finite dimensional subspaces is still injective and makes no assumption of the involved constants.

Proposition 3.6 (Uniqueness of the minimizer). *Let K possess the FBI property. Then the Tikhonov functional T_α has a unique minimizer.*

Proof. Let $u^{\alpha,\varepsilon}$ be a minimizer of T_α and $M \geq T_\alpha(u^{\alpha,\varepsilon}) \geq 0$. Since K possesses the FBI property, according to [8, Lemma 3], there is a positive constant $c = c(M, u^{\alpha,\varepsilon}, K) > 0$ that only depends on M, $u^{\alpha,\varepsilon}$ and K such that

$$T_\alpha(v) \geq T_\alpha(u^{\alpha,\varepsilon}) + c\|v - u^{\alpha,\varepsilon}\|_{\ell^2}^2, \quad \text{for all } v \text{ with } T_\alpha(v) \leq M.$$

Assume that there is another minimizer $\tilde{u}^{\alpha,\varepsilon}$ of T_α, i.e. $T_\alpha(\tilde{u}^{\alpha,\varepsilon}) = T_\alpha(u^{\alpha,\varepsilon})$, hence $T_\alpha(\tilde{u}^{\alpha,\varepsilon}) \leq M$. Then

$$0 = T_\alpha(\tilde{u}^{\alpha,\varepsilon}) - T_\alpha(u^{\alpha,\varepsilon}) \geq c\,\|u^{\alpha,\varepsilon} - \tilde{u}^{\alpha,\varepsilon}\|_{\ell^2}^2.$$

Since $c > 0$, we end in $u^{\alpha,\varepsilon} = \tilde{u}^{\alpha,\varepsilon}$. □

Remark 3.7. To the end of this section we remark on properties of the minimum-$\|\cdot\|_{\ell^1}$ solution $u^\dagger$ of $Ku = g$. In general, the minimum-$\|\cdot\|_{\ell^1}$ solution $u^\dagger$ of $Ku = g$ neither needs to be finitely supported, nor needs to be unique. If we assume that there is a finitely supported solution $u^\diamond \in \ell^0$ of $Ku = g$, then the set of all solutions of $Ku = g$ is given by

$$u^\diamond + \ker K.$$

If K possesses the FBI property, then the solution $u^\diamond$ is the unique solution in ℓ^0, hence $\ker K \subset \ell^2 \setminus \ell^0$. However, in general $u^\diamond \in \ell^0$ is not a minimum-$\|\cdot\|_{\ell^1}$ solution. In the following we assume that K possesses the FBI property. We denote the unique solution of $Ku = g$ in ℓ^0 with $u^\diamond$ and a minimum-$\|\cdot\|_{\ell^1}$ solution of $Ku = g$ with $u^\dagger$.

3.3 Stability and convergence rates

Stability and convergence rates for the Tikhonov functional (3.2) have been deduced in [25, 41, 42, 65]. Unlike for the Besov space penalties in chapter 2, one cannot use the general convergence rates results from [12, 49, 86, 87] here straightforward in order to obtain convergence rates in norm. These results lead to convergence rates in the Bregman distance only. In section 2.3 we used the lemma 4.7 from [7] to come to an estimate in a Hilbert space norm. Since $p = 1$ here, one cannot use the same statements.

In the following we cite stability and convergence rates results with respect to a minimum-$\|\cdot\|_{\ell^1}$ solution from [41]. We begin with the statement that the ℓ^1-penalized Tikhonov functional yields a regularization method.

Theorem 3.8 (Stability [41, proposition 7]). *Let K possess the FBI property. Assume that the operator equation $Ku = g$ attains a solution in ℓ^1 and that the parameter rule satisfies $\alpha \asymp \varepsilon$. Let $\varepsilon_k \to 0$, $k \to \infty$, and let $g^{\varepsilon_k} \in \mathcal{H}_2$ satisfy $\|g - g^{\varepsilon_k}\|_{\mathcal{H}_2} \leq \varepsilon_k$. Define the corresponding Tikhonov minimizers $u^{\alpha_k,\varepsilon_k}$ by*

$$u^{\alpha_k,\varepsilon_k} := \arg\min_{u \in \ell^2} \tfrac{1}{2}\|Ku - g^{\varepsilon_k}\|^2_{\mathcal{H}_2} + \alpha\|u\|_{\ell^1}.$$

Then there exists a minimum-$\|\cdot\|_{\ell^1}$ solution $u^\dagger$ of $Ku = g$ and a subsequence $(u^{\alpha_{k_j},\varepsilon_{k_j}})_{j\in\mathbb{N}}$ with $\|u^{\alpha_{k_j},\varepsilon_{k_j}} - u^\dagger\|_{\ell^1} \to 0$, $j \to \infty$. If $u^\dagger$ is unique, then $(u^{\alpha_k,\varepsilon_k})_{k\in\mathbb{N}}$ converges to $u^\dagger$ with respect to ℓ^1, i.e.

$$\|u^{\alpha_k,\varepsilon_k} - u^\dagger\|_{\ell^1} \to 0, \quad k \to \infty. \tag{3.8}$$

If additionally the minimum-$\|\cdot\|_{\ell^1}$ solution $u^\dagger$ is finitely supported and a certain source condition is satisfied, then even a statement of the rate of convergence is possible.

Theorem 3.9 (Convergence rate [41, theorem 15]). *Let K possess the FBI property, $u^\dagger \in \ell^0$ be a minimum-$\|\cdot\|_{\ell^1}$ solution of $Ku = g$, and $\|g - g^\varepsilon\|_{\mathcal{H}_2} \leq \varepsilon$. Moreover, let the following source condition be fulfilled:*

$$\operatorname{rg} K^* \cap \operatorname{Sign}(u^\dagger) \neq \emptyset. \tag{3.9}$$

Then for the minimizer $u^{\alpha,\varepsilon}$ of

$$T_\alpha(u) = \tfrac{1}{2}\|Ku - g^\varepsilon\|^2_{\mathcal{H}_2} + \alpha\|u\|_{\ell^1}$$

and the parameter rule $\alpha \asymp \varepsilon$ we get the convergence rate

$$\|u^{\alpha,\varepsilon} - u^\dagger\|_{\ell^1} = \mathcal{O}(\varepsilon). \tag{3.10}$$

Remark 3.10. To achieve the linear convergence rate (3.10), the source condition (3.9) is even necessary, cf. [42].

3.4 Beyond convergence rates: exact recovery

In the last paragraph we have seen that for a suitable a priori parameter rule the ℓ^1-penalized Tikhonov regularization yields a regularization method. If additionally the minimum-$\|\cdot\|_{\ell^1}$ solution $u^\dagger$ is finitely supported and the source condition (3.9) is fulfilled, then the convergence can be obtained with a linear rate.

In this paragraph we go beyond this question. We give an a priori parameter rule which ensures that the unknown support of the sparse solution $u^\diamond \in \ell^0$ is recovered exactly, i.e.

$$\operatorname{supp}(u^{\alpha,\varepsilon}) = \operatorname{supp}(u^\diamond).$$

We assume that K possesses the FBI property, and hence $u^\diamond$ is the unique solution of $Ku = g$ in ℓ^0. With I we denote the support of $u^\diamond$, i.e.

$$I := \operatorname{supp}(u^\diamond) := \{i \in \mathbb{Z} \,|\, u_i^\diamond \neq 0\}.$$

The results presented here are a generalization of [36,37] and [102,103]. In [36], Fuchs gives a condition in a finite dimensional setting, which ensures that the support of

$$u^{\alpha,0} := \operatorname*{arg\,min}_{u \in \ell^2} \tfrac{1}{2}\|Ku - g\|^2_{\mathcal{H}_2} + \alpha \|u\|_{\ell^1},$$

i.e. the Tikhonov minimizer with noiseless data g, and the support of $u^\diamond \in \ell^0$ coincide. In [37], he transfers his results from [36] to noisy signals. He derives a condition which ensures $\operatorname{supp}(u^{\alpha,\varepsilon}) = \operatorname{supp}(u^\diamond)$ for $\varepsilon > 0$, assuming that the so-called coherence parameter μ is small. It is defined by

$$\mu := \sup_{i \neq j} |\langle Ke_i, Ke_j \rangle|.$$

In inverse problems the ill-posedness of the operator K typically causes that two distinct images Ke_i and Ke_j, $i \neq j$, look much alike. Therefore, the coherence parameter μ becomes huge and Fuchs' results cannot be transfered to ill-posed inverse problems.

Independently, in [102,103] Tropp gives a condition which ensures exact recovery in terms of the ℓ^2 best approximation over the true support I. The proof techniques from [102,103] can be transfered to achieve an a priori parameter rule which ensures $\operatorname{supp}(u^{\alpha,\varepsilon}) = \operatorname{supp}(u^\diamond)$ on a certain condition.

3.4.1 Exact recovery for exact data

Before we investigate the a priori parameter rule, we cite Fuchs' recovery condition from [36] for exact data g. We slightly reformulate it, namely, in an infinite dimensional setting with K possessing the FBI property. The proof is contained in [36], however, we include it here due to the conceptual differences.

For the first statement we need the following notations. For a subset $J \subset \mathbb{Z}$, we denote with $P_J : \ell^2 \to \ell^2$ the projection onto J, i.e.

$$P_J u := \textstyle\sum_{j\in J} u_j e_j,$$

i.e. the coefficients $j \notin J$ are set to 0 and hence $\operatorname{supp}(P_J u) \subset J$. Moreover, for a finite dimensional linear operator B, with $B^\dagger$ we denote the pseudoinverse operator.

Theorem 3.11 (Exact recovery condition for exact data [36]). *Let $u^\diamond \in \ell^0$ with* $\operatorname{supp}(u^\diamond) = I$, *$g = Ku^\diamond$ the noiseless signal, and let K possess the FBI property. Moreover, let the* exact recovery condition (ERC) *be fulfilled:*

$$\sup_{i\in I^\complement} \|(KP_I)^\dagger K e_i\|_{\ell^1} < 1. \tag{3.11}$$

Then there exists a constant $\alpha_0 > 0$, such that for all $0 < \alpha < \alpha_0$ the support of $u^\diamond$ and the Tikhonov minimizer $u^{\alpha,0}$ coincide, where

$$u^{\alpha,0} := \operatorname*{arg\,min}_{u\in\ell^2} \tfrac{1}{2}\|Ku - g\|^2_{\mathcal{H}_2} + \alpha\|u\|_{\ell^1}.$$

Proof. To $u^\diamond \in \ell^0$ with $\operatorname{supp}(u^\diamond) = I$, the regularization parameter $\alpha > 0$ and exact data $g = Ku^\diamond$, we define u^* with $\operatorname{supp}(u^*) \subset \operatorname{supp}(u^\diamond)$ via $P_{I^\complement} u^* := 0$ and

$$u^* := (KP_I)^\dagger g - \alpha\,(P_I K^* K P_I)^{-1} \operatorname{sign}(u^\diamond).$$

Note that the operator $P_I K^* K P_I : \ell^2 \to \ell^2$ is a finite dimensional operator and that, for $u \in \ell^2$, it can be represented by

$$\begin{aligned} P_I K^* K P_I u = P_I K^* \big[\sum_{i \in I} u_i K e_i \big] &= \sum_{j \in I} \sum_{i \in I} u_i \langle K^* K e_i, e_j \rangle e_j \\ &= \sum_{j \in I} \sum_{i \in I} u_i \langle K e_i, K e_j \rangle e_j. \end{aligned}$$

Since $(KP_I)^\dagger g = u^\diamond$, we see that for sufficiently small $\alpha > 0$ and for all $i \in I$ it hold that

$$\big| \big((KP_I)^\dagger g \big)_i \big| = \big| (u^\diamond)_i \big| > \alpha \big| \big((P_I K^* K P_I)^{-1} \operatorname{sign}(u^\diamond) \big)_i \big|.$$

Hence $\operatorname{sign}(u^*) = \operatorname{sign}(u^\diamond)$ and in particular

$$\operatorname{supp}(u^*) = \operatorname{supp}(u^\diamond).$$

Next, we prove that u^* is the minimizer of the Tikhonov functional $\frac{1}{2} \| K \cdot - g \|^2_{\mathcal{H}_2} + \alpha \| \cdot \|_{\ell^1}$. Since we work with exact data $g = K u^\diamond$ and K possesses the FBI property, we see that

$$\begin{aligned} K u^* - K (KP_I)^\dagger g = K u^* - g &= -\alpha \, K P_I (P_I K^* K P_I)^{-1} \operatorname{sign}(u^\diamond) \\ &= -\alpha \, (P_I K^*)^\dagger \operatorname{sign}(u^\diamond). \end{aligned} \tag{3.12}$$

With that and since $P_I K^* (P_I K^*)^\dagger = (P_I K^* K P_I)\,(P_I K^* K P_I)^{-1} = P_I$ the condition (3.7.a) is fulfilled for J equal to the support I of $u^\diamond$, i.e.

$$P_I K^* (K u^* - g) = -\alpha \operatorname{sign}(u^\diamond).$$

The condition (3.7.b) is true with $J^\complement = I^\complement$. Here, again with equation (3.12), we get

$$P_{I^\complement} K^* (K u^* - g) = -\alpha \, P_{I^\complement} K^* (P_I K^*)^\dagger \operatorname{sign}(u^\diamond).$$

Hence, for all $j \in I^\complement$, we get

$$\begin{aligned} \big| \big(K^* (K u^* - g) \big)_j \big| &= \alpha \, \big| \big(K^* (P_I K^*)^\dagger \operatorname{sign}(u^\diamond)) \big)_j \big| \\ &= \alpha \, \Big| \big\langle e_j, K^* (P_I K^*)^\dagger \operatorname{sign}(u^\diamond) \big\rangle \Big| \\ &= \alpha \, \Big| \big\langle (KP_I)^\dagger K e_j, \operatorname{sign}(u^\diamond) \big\rangle \Big|. \end{aligned}$$

By assumption (3.11) this is smaller than α, since for $j \in I^{\complement}$ it holds that

$$\left|\langle (KP_I)^\dagger Ke_j, \operatorname{sign}(u^\diamond)\rangle\right| \leq \sup_{i\in I^{\complement}} \sup_{\operatorname{supp}(u)=I} \left|\langle (KP_I)^\dagger Ke_i, \operatorname{sign}(u)\rangle\right|$$

$$= \sup_{i\in I^{\complement}} \sum_{k\in I} \left|\left((KP_I)^\dagger Ke_i\right)_k\right| = \sup_{i\in I^{\complement}} \|(KP_I)^\dagger Ke_i\|_{\ell^1} < 1,$$

since $\operatorname{supp}\left((KP_I)^\dagger Ke_i\right) \subset I$. Since K has the FBI property, $u^* = u^{\alpha,0}$ is the unique minimizer. □

3.4.2 Exact recovery in the presence of noise

In [37], Fuchs transfers his results from [36] to noisy signals. He derives a condition for exact recovery in terms of the coherence parameter μ. Typically, for ill-posed inverse problems the coherence parameter μ gets huge and hence these results cannot be used.

Independently, in [102, 103] Tropp gives a condition in a finite dimensional setting, which ensures exact recovery and that works without the concept of coherence. His condition depends on the ℓ^2 best approximation over the true support I, i.e. on the term

$$P_{\operatorname{rg}(KP_I)} g^\varepsilon.$$

We slightly modify Tropp's condition and come to an a priori parameter rule that ensures exact recovery. To prove it, we use the techniques from [102, 103].

Assume that instead of exact data $g = Ku^\diamond \in \mathcal{H}_2$, only a noisy version

$$g^\varepsilon = g + \eta = Ku^\diamond + \eta$$

with noise level $\|g^\varepsilon - g\|_{\mathcal{H}_2} = \|\eta\|_{\mathcal{H}_2} \leq \varepsilon$ can be observed. In this case we choose the regularization parameter α in dependence of the noise η.

Theorem 3.12 (Lower bound on α). *Let $u^\diamond \in \ell^0$, $\operatorname{supp}(u^\diamond) = I$, and $g^\varepsilon = Ku^\diamond + \eta$ the noisy data with noise level $\|\eta\|_{\mathcal{H}_2} \leq \varepsilon$. Assume that the operator K is bounded by 1, i.e. $\|K\|_{\ell^2,\mathcal{H}_2} \leq 1$, and that K possesses the FBI property. If the following condition holds,*

$$\sup_{i\in I^{\complement}} \|(KP_I)^\dagger Ke_i\|_{\ell^1} < 1, \tag{3.13}$$

then the parameter rule

$$\alpha > \frac{1 + \sup_{i\in I^{\complement}} \|(KP_I)^\dagger Ke_i\|_{\ell^1}}{1 - \sup_{i\in I^{\complement}} \|(KP_I)^\dagger Ke_i\|_{\ell^1}} \sup_{i\in\mathbb{Z}} |\langle \eta, Ke_i\rangle| \tag{3.14}$$

ensures that the support of $u^{\alpha,\varepsilon} := \arg\min T_\alpha(u)$ *is contained in* I.

Proof. Let $u_I^{\alpha,\varepsilon}$ be the unique minimizer of T_α with coefficients supported on I, i.e. $u_I^{\alpha,\varepsilon}$ is defined by

$$u_I^{\alpha,\varepsilon} := \arg\min_{\operatorname{supp}(u)\subset I} T_\alpha(u). \tag{3.15}$$

Provided that (3.14) holds, we show that the value of the Tikhonov functional T_α increases, if we add an arbitrarily small $v \in \ell^2$, i.e.

$$T_\alpha(u_I^{\alpha,\varepsilon} + v) - T_\alpha(u_I^{\alpha,\varepsilon}) > 0.$$

This proves the claim, because the Tikhonov functional T_α is convex. For $v \in \ell^2 \setminus \ell^1$, this is obvious, since in this case $T_\alpha(u_I^{\alpha,\varepsilon} + v) = \infty$. Thus, let $v \in \ell^1$. Splitting v into the components $v = v_I + v_{I^\complement}$ with $\operatorname{supp}(v_I) \subset I$ and $\operatorname{supp}(v_{I^\complement}) \subset I^\complement$ allows to reformulate

$$\begin{aligned} T_\alpha(u_I^{\alpha,\varepsilon} + v) - T_\alpha(u_I^{\alpha,\varepsilon}) &= T_\alpha(u_I^{\alpha,\varepsilon} + v_I) - T_\alpha(u_I^{\alpha,\varepsilon}) + \tfrac{1}{2}\|Kv_{I^\complement}\|_{\mathcal{H}_2}^2 \\ &+ \operatorname{Re}\langle Ku_I^{\alpha,\varepsilon} - g^\varepsilon, Kv_{I^\complement}\rangle + \operatorname{Re}\langle Kv_I, Kv_{I^\complement}\rangle + \alpha\|v_{I^\complement}\|_{\ell^1}. \end{aligned}$$

Since $u_I^{\alpha,\varepsilon}$ is the unique minimizer of T_α with coefficients supported on I and $\|Kv_{I^\complement}\|_{\mathcal{H}_2}^2 \geq 0$ we can estimate

$$\begin{aligned} &T_\alpha(u_I^{\alpha,\varepsilon} + v) - T_\alpha(u_I^{\alpha,\varepsilon}) \\ &\quad\geq \alpha\|v_{I^\complement}\|_{\ell^1} - |\langle Ku_I^{\alpha,\varepsilon} - g^\varepsilon, Kv_{I^\complement}\rangle| - |\langle Kv_I, Kv_{I^\complement}\rangle|. \end{aligned} \tag{3.16}$$

With $Kv_{I^\complement} = \sum_{i\in I^\complement} v_i Ke_i$, using the linearity of the inner product $\langle\cdot,\cdot\rangle$, and Hölder's inequality we get

$$\begin{aligned} |\langle Ku_I^{\alpha,\varepsilon} - g^\varepsilon, Kv_{I^\complement}\rangle| &\leq \sum_{i\in I^\complement} |v_i|\, |\langle Ku_I^{\alpha,\varepsilon} - g^\varepsilon, Ke_i\rangle| \\ &\leq \|v_{I^\complement}\|_{\ell^1} \sup_{i\in I^\complement} |\langle Ku_I^{\alpha,\varepsilon} - g^\varepsilon, Ke_i\rangle|. \end{aligned}$$

With Hölder's inequality the second inner product of (3.16) can be estimated by

$$|\langle Kv_I, Kv_{I^\complement}\rangle| = |\langle K^*Kv_I, v_{I^\complement}\rangle| \leq \|K^*Kv_I\|_{\ell^\infty}\, \|v_{I^\complement}\|_{\ell^1}.$$

Plugging both estimates into the above inequality (3.16), we end in

$$\begin{aligned} &T_\alpha(u_I^{\alpha,\varepsilon} + v) - T_\alpha(u_I^{\alpha,\varepsilon}) \\ &\quad\geq \|v_{I^\complement}\|_{\ell^1} \left(\alpha - \sup_{i\in I^\complement} |\langle Ku_I^{\alpha,\varepsilon} - g^\varepsilon, Ke_i\rangle| - \|K^*Kv_I\|_{\ell^\infty}\right). \end{aligned}$$

Since K is a bounded operator and since we choose v arbitrarily small, hence $\|K^*Kv_I\|_{\ell^\infty}$ is arbitrarily small, the restricted minimizer $u_I^{\alpha,\varepsilon}$ of T_α is also the global minimizer if

$$\alpha > \sup_{i\in I^{\complement}} |\langle Ku_I^{\alpha,\varepsilon} - g^\varepsilon, Ke_i\rangle|. \tag{3.17}$$

Define $J := \operatorname{supp}(u_I^{\alpha,\varepsilon}) \subset I$ and $J^{\complement} := I \setminus J$. From the optimality condition (3.7) of the restricted optimization problem (3.15) we know

$$\begin{aligned} -(K^*(Ku_I^{\alpha,\varepsilon} - g^\varepsilon))_j &= \alpha \operatorname{sign}\big((u_I^{\alpha,\varepsilon})_j\big), & j &\in J,\\ |K^*(Ku_I^{\alpha,\varepsilon} - g^\varepsilon)|_j &\le \alpha, & j &\in J^{\complement}. \end{aligned}$$

Hence there is a $w \in \ell^\infty$, with $\|w\|_{\ell^\infty} \le 1$ and which is supported by I, with

$$P_I K^*(Ku_I^{\alpha,\varepsilon} - g^\varepsilon) = \alpha\, w.$$

Multiplying both sides by $(P_I K^* K P_I)^{-1}$, we see its equivalence to

$$u_I^{\alpha,\varepsilon} - (KP_I)^\dagger g^\varepsilon = \alpha (P_I K^* K P_I)^{-1} w,$$

hence

$$\begin{aligned} Ku_I^{\alpha,\varepsilon} - g^\varepsilon &= Ku_I^{\alpha,\varepsilon} - KP_I(KP_I)^\dagger g^\varepsilon + KP_I(KP_I)^\dagger g^\varepsilon - g^\varepsilon\\ &= \alpha (P_I K^*)^\dagger w + \big(KP_I(KP_I)^\dagger - \mathrm{Id}\big)(g^\varepsilon). \end{aligned}$$

Since $KP_I(KP_I)^\dagger - \mathrm{Id}$ equals the orthogonal projection on $\operatorname{rg}(KP_I)^\perp$ and $g = Ku^\diamond \in \operatorname{rg}(KP_I)$, we get

$$\begin{aligned} Ku_I^{\alpha,\varepsilon} - g^\varepsilon &= \alpha (P_I K^*)^\dagger w + \big(KP_I(KP_I)^\dagger - \mathrm{Id}\big)(g^\varepsilon - g)\\ &= \alpha (P_I K^*)^\dagger w + \big(KP_I(KP_I)^\dagger - \mathrm{Id}\big)(\eta). \end{aligned}$$

With that we continue with condition (3.17) and estimate for $i \in I^{\complement}$

$$\begin{aligned} &|\langle Ku_I^{\alpha,\varepsilon} - g^\varepsilon, Ke_i\rangle|\\ &\quad\le \alpha |\langle w, (KP_I)^\dagger Ke_i\rangle| + |\langle (KP_I(KP_I)^\dagger - \mathrm{Id})\eta, Ke_i\rangle|\\ &\quad\le \alpha \|(KP_I)^\dagger Ke_i\|_{\ell^1} + |\langle \eta, (KP_I(KP_I)^\dagger - \mathrm{Id})Ke_i\rangle|, \end{aligned}$$

where we used Hölder's inequality and $\|w\|_{\ell^\infty} \le 1$.

The expression $|\langle \eta, (KP_I(KP_I)^\dagger - \mathrm{Id})Ke_i\rangle|$ is challengingly touchable. With Hölder's inequality we get an upper bound that is easier to use, namely, for $i \in I^\complement$ it holds that

$$\begin{aligned}|\langle \eta, (\mathrm{Id} - KP_I(KP_I)^\dagger)Ke_i\rangle| &= |\langle K^*\eta, (\mathrm{Id} - (KP_I)^\dagger K)e_i\rangle_{\ell^2}| \\ &\le \|K^*\eta\|_{\ell^\infty} \, \|(\mathrm{Id} - (KP_I)^\dagger K)e_i\|_{\ell^1}.\end{aligned}$$

Since $i \in I^\complement$ and $\mathrm{supp}((KP_I)^\dagger Ke_i) \subset I$ we get

$$\sup_{i \in I^\complement} \|e_i - (KP_I)^\dagger Ke_i\|_{\ell^1} = 1 + \sup_{i \in I^\complement} \|(KP_I)^\dagger Ke_i\|_{\ell^1}.$$

Hence we end in the estimation

$$\sup_{i \in I^\complement} |\langle \eta, P_{\mathrm{rg}(KP_I)^\perp} Ke_i\rangle| \le \sup_{i \in \mathbb{Z}} |\langle \eta, Ke_i\rangle| \left(1 + \sup_{i \in I^\complement} \|(KP_I)^\dagger Ke_i\|_{\ell^1}\right). \tag{3.18}$$

Thus the condition (3.13) together with the parameter choice rule (3.14) ensures that the support is contained in I. □

Remark 3.13. We remark, that the assumption $\|K\|_{\ell^2,\mathcal{H}_2} \le 1$ has not been used in the proof of theorem 3.12. It has been introduced for the sake of notational simplification, however, we will need it for some estimations afterwards. The assumption that K is an operator with norm bounded by 1 is a common condition for Tikhonov functionals with sparsity constraints, cf. e.g. [25]. Actually, this is not really a limitation, because every inverse problem $Ku = g$ with bounded operator K can be rescaled so that $\|K\|_{\ell^2,\mathcal{H}_2} \le 1$.

Remark 3.14. Instead of using the estimation (3.18), one can alternatively use another upper bound for $|\langle \eta, (KP_I(KP_I)^\dagger - \mathrm{Id})Ke_i\rangle|$. Remember that $KP_I(KP_I)^\dagger - \mathrm{Id}$ is the orthogonal projection on $\mathrm{rg}(KP_I)^\perp$. Then, since K and orthogonal projections are bounded by 1, we can estimate for $i \in I^\complement$ with the Cauchy-Schwarz inequality as follows

$$\begin{aligned}|\langle \eta, (KP_I(KP_I)^\dagger - \mathrm{Id})Ke_i\rangle| &\le \|\eta\|_{\mathcal{H}_2} \|(KP_I(KP_I)^\dagger - \mathrm{Id})Ke_i\|_{\mathcal{H}_2} \\ &\le \varepsilon.\end{aligned} \tag{3.19}$$

In general, one cannot say which estimate gives a sharper bound, inequality (3.18) or inequality (3.19). However, in praxis the noise η often is uniformly distributed and hence $\sup_{i\in\mathbb{Z}} |\langle \eta, Ke_i\rangle| \ll \varepsilon$ holds. In this case the estimation with Hölder's inequality (3.18) gives a sharper estimation. We use (3.18) for the examples from mass spectrometry and digital holography in chapter 5.

Theorem 3.12 gives a lower bound on the regularization parameter α to ensure $\operatorname{supp}(u^{\alpha,\varepsilon}) \subset \operatorname{supp}(u^\diamond)$. To guarantee $\operatorname{supp}(u^{\alpha,\varepsilon}) = \operatorname{supp}(u^\diamond)$ we need an additional upper bound for α. The following corollary leads to that purpose.

Corollary 3.15 (Error estimate)**.** *Let the assumptions of theorem 3.12 hold, i.e.* $\operatorname{supp}(u^{\alpha,\varepsilon}) \subset I$. *Then the following error estimate holds:*

$$\|u^\diamond - u^{\alpha,\varepsilon}\|_{\ell^\infty} \le (\alpha + \sup_{i\in\mathbb{Z}} |\langle \eta, Ke_i\rangle|) \|(P_I K^* K P_I)^{-1}\|_{\ell^1,\ell^1}. \qquad (3.20)$$

Proof. From the assumptions of theorem 3.12 we have $\operatorname{supp}(u^{\alpha,\varepsilon}) \subset \operatorname{supp}(u^\diamond)$. From the optimality condition (3.5) we know that for $u^{\alpha,\varepsilon}$ there is a $w \in \ell^\infty$ with $\|w\|_{\ell^\infty} \le 1$ such that

$$-K^*(Ku^{\alpha,\varepsilon} - g^\varepsilon) = \alpha\, w.$$

Hence it holds that

$$\begin{aligned} -P_I K^* K P_I (u^{\alpha,\varepsilon} - u^\diamond) &= -P_I K^* (Ku^{\alpha,\varepsilon} - g) \\ &= -P_I K^* (Ku^{\alpha,\varepsilon} - g^\varepsilon) - P_I K^* \eta \\ &= \alpha P_I w - P_I K^* \eta. \end{aligned}$$

Since $\|w\|_{\ell^\infty} \le 1$, with Hölder's inequality we can estimate for all $j \in I$

$$\begin{aligned} |(u^{\alpha,\varepsilon} - u^\diamond)_j| &= |\langle u^{\alpha,\varepsilon} - u^\diamond, e_j\rangle| \\ &= \big|\langle (P_I K^* K P_I)^{-1}(\alpha P_I w - P_I K^* \eta), e_j\rangle\big| \\ &\le \alpha |\langle P_I w, (P_I K^* K P_I)^{-1} e_j\rangle| + |\langle P_I K^*\eta, (P_I K^* K P_I)^{-1} e_j\rangle| \\ &\le (\alpha \|w\|_{\ell^\infty} + \|P_I K^* \eta\|_{\ell^\infty}) \|(P_I K^* K P_I)^{-1}\|_{\ell^1,\ell^1} \\ &\le (\alpha + \sup_{i\in\mathbb{Z}} |\langle \eta, Ke_i\rangle|) \|(P_I K^* K P_I)^{-1}\|_{\ell^1,\ell^1}. \end{aligned}$$

□

The error estimate (3.20) yields an upper bound for the regularization parameter α, namely,

$$\alpha < \frac{\min_{i\in I} |u_i^\diamond|}{\|(P_I K^* K P_I)^{-1}\|_{\ell^1,\ell^1}} - \sup_{i\in\mathbb{Z}} |\langle \eta, Ke_i\rangle|. \qquad (3.21)$$

Together with the lower bound (3.14), equation (3.21) ensures exact recovery. Note that the interval of convenient regularization parameters resulting from (3.14) and (3.21) depends on the noise-to-signal ratio. For heavily disturbed signals it can be empty. The following theorem gives a sufficient condition for the existence of a regularization parameter α which provides exact recovery. Due to corollary 3.15, equation (3.20), the regularization parameter has to be chosen as small as possible.

Theorem 3.16 (Exact recovery condition in the presence of noise). *Let $u^\diamond \in \ell^0$ with $\operatorname{supp}(u^\diamond) = I$ and $g^\varepsilon = Ku^\diamond + \eta$ the noisy data with noise level $\|\eta\|_{\mathcal{H}_2} \leq \varepsilon$ and noise-to-signal ratio*

$$r_{\varepsilon/u} := \frac{\sup_{i\in\mathbb{Z}} |\langle \eta, Ke_i\rangle|}{\min_{i\in I} |u_i^\diamond|}.$$

Assume that the operator K is bounded by 1 and that K possesses the FBI property. Then the exact recovery condition in the presence of noise (εERC)

$$\sup_{i\in I^{\complement}} \|(KP_I)^\dagger Ke_i\|_{\ell^1} < 1 - 2r_{\varepsilon/u}\|(P_IK^*KP_I)^{-1}\|_{\ell^1,\ell^1} \tag{3.22}$$

ensures that there is a suitable regularization parameter α,

$$\frac{1+\sup_{i\in I^{\complement}}\|(KP_I)^\dagger Ke_i\|_{\ell^1}}{1-\sup_{i\in I^{\complement}}\|(KP_I)^\dagger Ke_i\|_{\ell^1}} \sup_{i\in\mathbb{Z}} |\langle \eta, Ke_i\rangle| < \alpha \tag{3.23}$$

$$\alpha < \frac{\min_{i\in I}|u_i^\diamond|}{\|(P_IK^*KP_I)^{-1}\|_{\ell^1,\ell^1}} - \sup_{i\in\mathbb{Z}} |\langle \eta, Ke_i\rangle|,$$

which provides exact recovery of I, i.e. the support of the minimizer $u^{\alpha,\varepsilon} := \arg\min T_\alpha(u)$ coincides with $\operatorname{supp}(u^\diamond) = I$.

In fact, the parameter choice rule (3.23) is not an a priori parameter rule $\alpha = \alpha(\varepsilon)$, since it depends on the noise η. However, the term $\sup_{i\in\mathbb{Z}} |\langle \eta, Ke_i\rangle|$ is related to the noise level and it can be estimated by ε, cf. remark 3.14. Further notice, that for a small perturbation η the εERC (3.22) approximates the ERC (3.11).

Due to the expressions

$$\|(P_IK^*KP_I)^{-1}\|_{\ell^1,\ell^1} \quad \text{and} \quad \sup_{i\in I^{\complement}} \|(KP_I)^\dagger Ke_i\|_{\ell^1},$$

the εERC (3.22) is hard to evaluate, especially since the support I is unknown. Therefore we give a weaker sufficient recovery condition, that depends on inner products of images of K restricted to I and $I^{\complement}$. For the sake of an easier presentation we define

$$\mathrm{COR}_I := \sup_{i \in I} \sum_{\substack{j \in I \\ j \neq i}} |\langle Ke_i, Ke_j \rangle|, \tag{3.24}$$

$$\mathrm{COR}_{I^{\complement}} := \sup_{i \in I^{\complement}} \sum_{j \in I} |\langle Ke_i, Ke_j \rangle|. \tag{3.25}$$

Proposition 3.17 (Neumann exact recovery condition). *Let* $u^\diamond \in \ell^0$ *with* $\operatorname{supp}(u^\diamond) = I$ *and* $g^\varepsilon = Ku^\diamond + \eta$ *the noisy data with noise level* $\|\eta\|_{\mathcal{H}_2} \le \varepsilon$ *and noise-to-signal ratio* $r_{\varepsilon/u}$. *Assume that the operator* K *is bounded by 1 and that* K *possesses the FBI property. Then the* Neumann exact recovery condition in the presence of noise (Neumann εERC)

$$\mathrm{COR}_I + \mathrm{COR}_{I^{\complement}} < \min_{i \in I} \|Ke_i\|^2_{\mathcal{H}_2} - 2r_{\varepsilon/u} \tag{3.26}$$

ensures that there is a suitable regularization parameter α,

$$\frac{\min_{i \in I} \|Ke_i\|^2_{\mathcal{H}_2} - \mathrm{COR}_I + \mathrm{COR}_{I^{\complement}}}{\min_{i \in I} \|Ke_i\|^2_{\mathcal{H}_2} - \mathrm{COR}_I - \mathrm{COR}_{I^{\complement}}} \sup_{i \in \mathbb{Z}} |\langle \eta, Ke_i \rangle| < \alpha \tag{3.27}$$
$$\alpha < \big(\min_{i \in I} \|Ke_i\|^2_{\mathcal{H}_2} - \mathrm{COR}_I \big) \min_{i \in I} |u_i^\diamond| - \sup_{i \in \mathbb{Z}} |\langle \eta, Ke_i \rangle|,$$

which provides exact recovery of I, *i.e. the support of* $u^{\alpha,\varepsilon}$ *coincides with* $\operatorname{supp}(u^\diamond) = I$.

Proof. For the deduction of conditions (3.26) and (3.27) from conditions (3.22) and (3.23), respectively, we use

$$\sup_{i \in I^{\complement}} \|(KP_I)^\dagger Ke_i\|_{\ell^1} \le \|(P_I K^* K P_I)^{-1}\|_{\ell^1,\ell^1} \sup_{i \in I^{\complement}} \|P_I K^* K e_i\|_{\ell^1}.$$

The second term we can rewrite as follows

$$\sup_{i \in I^{\complement}} \|P_I K^* K e_i\|_{\ell^1} = \sup_{i \in I^{\complement}} \sum_{j \in I} |\langle Ke_i, Ke_j \rangle| = \mathrm{COR}_{I^{\complement}}.$$

For the first term, we split the operator $P_I K^* K P_I$ into diagonal M_{diag} and off-diagonal M_{off} and use Neumann series expansion for its inverse,

$$(P_I K^* K P_I)^{-1} = (M_{\mathrm{diag}} + M_{\mathrm{off}})^{-1} = \big(\mathrm{Id} - ((\mathrm{Id} - M_{\mathrm{diag}}) - M_{\mathrm{off}})\big)^{-1}.$$

With (3.26) it holds that $\mathrm{COR}_I < \min_{i\in I} \|Ke_i\|^2_{\mathcal{H}_2}$. With that and since $\|K\|_{\ell^2,\mathcal{H}_2} \leq 1$ we get

$$\begin{aligned}
\|P_I K^* K P_I - \operatorname{Id}\|_{\ell^1,\ell^1} &\leq \|\operatorname{Id} - M_{\text{diag}}\|_{\ell^1,\ell^1} + \|M_{\text{off}}\|_{\ell^1,\ell^1} \\
&= 1 - \min_{i\in I} \|Ke_i\|^2_{\mathcal{H}_2} + \sup_{i\in I} \sum_{\substack{j\in I \\ j\neq i}} |\langle Ke_i, Ke_j\rangle| \\
&= 1 - \min_{i\in I} \|Ke_i\|^2_{\mathcal{H}_2} + \mathrm{COR}_I < 1.
\end{aligned}$$

Hence with Neumann series estimate we get

$$\begin{aligned}
\|(P_I K^* K P_I)^{-1}\|_{\ell^1,\ell^1} &\leq \frac{1}{1 - \|P_I K^* K P_I - \operatorname{Id}\|_{\ell^1,\ell^1}} \\
&\leq \frac{1}{\min\limits_{i\in I} \|Ke_i\|^2 - \mathrm{COR}_I}.
\end{aligned}$$

With these two estimates the proposition follows. □

Remark 3.18. By the assumptions of proposition 3.17, the operator K is bounded by 1, i.e. $\|Ke_i\|_{\mathcal{H}_2} \leq 1$ for all $i \in \mathbb{Z}$. Hence to ensure the Neumann εERC (3.26), one has necessarily for the noise-to-signal ratio $r_{\varepsilon/u} < 1/2$. For a lot of examples one can normalize K, so that $\|Ke_i\|_{\mathcal{H}_2} = 1$ holds for all $i \in \mathbb{Z}$. We do this for the examples from mass spectrometry and digital holography in chapter 5. In this case the Neumann εERC (3.26) reads as

$$\mathrm{COR}_I + \mathrm{COR}_{I^\complement} < 1 - 2r_{\varepsilon/u}.$$

This condition coincides with the OMP results presented afterwards in chapter 4.

Remark 3.19 (Exact recovery condition in terms of coherence)**.** As already mentioned, for sparse approximation problems the behavior of the Tikhonov minimizer is often characterized with the so-called *coherence parameter* μ, which is defined by

$$\mu := \sup_{i\neq j} |\langle Ke_i, Ke_j\rangle|.$$

Sometimes the so-called *cumulative coherence* μ_1 is used, too, which for a positive integer m is defined by

$$\mu_1(m) := \sup_{\substack{\Lambda\subset\mathbb{Z} \\ |\Lambda|=m}} \sup_{i\notin\Lambda} \sum_{j\in\Lambda} |\langle Ke_i, Ke_j\rangle|.$$

Since the exact recovery conditions in terms of the coherence parameter μ and the cumulative coherence μ_1 are rough, we abstain from formulating them here. We just remark that the correlations COR_I and $\mathrm{COR}_{I^\complement}$ can be estimated from above, with $N := \|u^\diamond\|_{\ell^0}$, by

$$\mathrm{COR}_I = \sup_{i\in I} \sum_{\substack{j\in I\\ j\neq i}} |\langle Ke_i, Ke_j\rangle| \leq \mu_1(N-1) \leq (N-1)\,\mu,$$

$$\mathrm{COR}_{I^\complement} = \sup_{i\in I^\complement} \sum_{j\in I} |\langle Ke_i, Ke_j\rangle| \leq \mu_1(N) \leq N\,\mu.$$

With that the exact recovery conditions in terms of the coherence parameter μ and the cumulative coherence μ_1 can easily be deduced from conditions (3.26) and (3.27).

3.5 Conclusion

With the papers [41] and [42], the analysis of a priori parameter rules for ℓ^1-penalized Tikhonov functionals seemed completed. On the common parameter rule $\alpha \asymp \varepsilon$, linear, i.e. best possible, convergence is guaranteed. In this chapter we have gone beyond this question by presenting an a priori parameter rule which ensures exact recovery of the unknown support of $u^\diamond \in \ell^0$.

Granted, to apply the Neumann εERC (3.26) and the Neumann parameter rule (3.27) one has to know the support I. However, with a certain prior knowledge the correlations

$$\mathrm{COR}_I := \sup_{i\in I} \sum_{\substack{j\in I\\ j\neq i}} |\langle Ke_i, Ke_j\rangle| \quad \text{and} \quad \mathrm{COR}_{I^\complement} := \sup_{i\in I^\complement} \sum_{j\in I} |\langle Ke_i, Ke_j\rangle|,$$

can be estimated from above a priori, especially when the support I is not known exactly. That way it is possible to obtain a priori computable conditions for exact recovery. We do this in chapter 5 exemplarily with impulse trains convolved with a Gaussian kernel and for characteristic functions convolved with a Fresnel function. This shows the practical relevance of the condition.

For the examples in chapter 5, a nonnegativity constraint is valid, i.e. one knows a priori that the coefficients $u_i^\diamond$ are nonnegative. Possibly, bringing in this additional prior knowledge may lead to tighter exact recovery conditions. We postpone this idea for future work.

Another further direction of research may be to analyze other types of Tikhonov functionals. In [111], the so-called elastic-net regularization is presented, where one minimizes the (strictly convex) Tikhonov-type functional

$$T_{\alpha,\beta}(u) = \frac{1}{2}\|Ku - g^{\varepsilon}\|^2_{\mathcal{H}_2} + \alpha\|u\|_{\ell^1} + \frac{\beta}{2}\|u\|^2_{\ell^2}.$$

The elastic-net functional encourages a grouping effect by the use of the additional ℓ^2 penalty. Furthermore, minimization algorithms sometimes outperform those for the ℓ^1-penalized functional, cf. e.g. [52]. The (unique) minimizer is sparse for all $\alpha > 0$ and $\beta > 0$. Interesting questions concerning exact recovery for the elastic-net functional are e.g. the following:

- For which parameters α and β does the support of the elastic-net minimizer coincide with the support of the real solution $u^{\diamond}$?
- For which parameter $\beta > 0$ (and on which additional conditions) does the support of the elastic-net minimizer and the ℓ^1-penalized Tikhonov minimizer coincide?

Some subquestion have already been answered in [92]. The other ideas we postpone for future work.

4 Greedy solution by means of the orthogonal matching pursuit

4.1 Introduction

In this chapter we make almost the same assumptions as in chapter 3 (Tikhonov regularization with an ℓ^1 penalty). We consider inverse problems with a bounded, injective, linear operator $A : \mathcal{B}_1 \to \mathcal{H}_2$ between the Banach space $\mathcal{B}_1$ and the Hilbert space $\mathcal{H}_2$,

$$Af = g. \tag{4.1}$$

Moreover, we assume that for an unknown $g \in \operatorname{rg} A$, we are given a noisy observation g^ε with $\|g - g^\varepsilon\|_{\mathcal{H}_2} \leq \varepsilon$ and try to reconstruct the solution of $Af = g$ from the knowledge of g^ε.

We are particularly interested in the case where the unknown solution $f^\diamond$ may be expressed sparsely. Different from chapter 3, we assume that $f^\diamond$ is not sparse in an orthonormal basis but sparse in a known dictionary. I.e. we consider that there is a family $\Psi := \{\psi_i\}_{i\in\mathbb{Z}} \subset \mathcal{B}_1$ of unit-normed vectors which span the space in which we expect the solution and which we call *dictionary*. With sparse we still mean that there exists a decomposition of $f^\diamond$ with a finite number of atoms $\psi_i \in \Psi$,

$$f^\diamond = \sum_{i\in\mathbb{Z}} u_i^\diamond \psi_i \quad \text{with} \quad u^\diamond \in \ell^2(\mathbb{Z}, \mathbb{R}), \quad \|u^\diamond\|_{\ell^0} < \infty.$$

The uncommon choice of the index set $\mathbb{Z}$ of the dictionary Ψ accounts for the examples in chapter 5. With I we still denote the support of

$u^\diamond$, i.e. $I := \operatorname{supp}(u^\diamond) := \{i \in \mathbb{Z} \,|\, u_i^\diamond \neq 0\}$. Since the dictionary atoms $\{\psi_i\}_{i\in\mathbb{Z}}$ need not necessarily be linearly independent, there may be more than one index set I so that f decomposes into atoms $\{\psi_i\}_{i\in I}$. Therefore, we assume I to be one of the smallest among all those index sets. Further, for any subset $J \subset \mathbb{Z}$ and any dictionary $\Theta = \{\vartheta_i\}_{i\in\mathbb{Z}}$ we denote the restricted dictionary by $\Theta(J) := \{\vartheta_i \,|\, i \in J\}$.

In the following, an approximate solution shall be found by deriving iteratively the maximal correlation between the residual and the unit-normed atoms of the dictionary

$$\Phi := \{\varphi_i\}_{i\in\mathbb{Z}} := \left\{ \frac{A\psi_i}{\|A\psi_i\|} \right\}_{i\in\mathbb{Z}}.$$

Since the operator A is injective, we get that $A\psi_i \neq 0$ for all $i \in \mathbb{Z}$, and hence the dictionary Φ is well defined. In contrast to chapter 3, we here define the elements φ_i to be unit-normed. The normalization of the atoms φ_i makes sense in a lot of applications, as e.g. in out-of-field particle detection in digital holography, cf. [98]. To find an approximate solution, we iteratively select that unit-normed atom from the dictionary Φ, which is mostly correlated with the residual, hence the name "greedy" method. To stabilize the solution the iteration has to be stopped early enough.

For solving the operator equation $Af = g$ with noiseless data g and the case where only noisy data g^ε with noise-bound $\|g - g^\varepsilon\|_{\mathcal{H}_2} \leq \varepsilon$ are available, we use the orthogonal matching pursuit (OMP). It was first proposed in the signal processing context by Davis et al. in [73] and Pati et al. in [83], as an improvement upon the matching pursuit algorithm [74]. For a comprehensive presentation of OMP see e.g. [72].

Algorithm 4.1 Orthogonal Matching Pursuit (OMP)

Set $k := 0$ and $I^0 := \emptyset$.
Initialize $r^0 := g^\varepsilon$ (resp. $r^0 := g$ for $\varepsilon = 0$) and $\widehat{f}^0 := 0$.
  while $\|r^k\|_{\mathcal{H}_2} > \varepsilon$ (resp. $\|r^k\|_{\mathcal{H}_2} \neq 0$) **do**
    $k := k + 1$,
   Greedy atom selection
    $i_k \in \arg\sup \left\{ |\langle r^{k-1}, \varphi_i \rangle| \,\middle|\, i \in \mathbb{Z} \right\}$,
    $I^k := I^{k-1} \cup \{i_k\}$,
   Project onto span $\Psi(I^k)$
    $\widehat{f}^k := \arg\min \left\{ \|g^\varepsilon - A\widehat{f}\|^2_{\mathcal{H}_2} \,\middle|\, \widehat{f} \in \operatorname{span} \Psi(I^k) \right\}$,
    $r^k := g^\varepsilon - A\widehat{f}^k$.
  end while

Remark that in infinite dimensional Hilbert spaces the supremum

$$\sup\{|\langle r^{k-1}, \varphi_i\rangle| \,|\, i \in \mathbb{Z}\} \tag{4.2}$$

does not have to be realized. Because of that, OMP has a variant—called weak orthogonal matching pursuit (WOMP)—which does not choose the optimal atom in the sense of (4.2) but only one that is nearly optimal. For some fixed $\omega \in (0,1]$, an index $i_k \in \mathbb{Z}$ is chosen, which fulfills

$$|\langle r^{k-1}, \varphi_{i_k}\rangle| \geq \omega \sup\{|\langle r^{k-1}, \varphi_i\rangle| \,|\, i \in \mathbb{Z}\}.$$

In comparison with minimization algorithms for the ℓ^1-penalized Tikhonov functional, as e.g. iterated soft- and hard-thresholding [7, 25] or miscellaneous active-set methods [44, 61], the orthogonal matching pursuit is a more efficient algorithm for solving the linear operator equation (4.1) with sparsity constraints. Since OMP never selects the same atom ψ_i twice, the signal possibly can be reconstructed after $N := \|u^\diamond\|_{\ell^0}$ steps. However, OMP in general is not a regularization method. There are even undisturbed signals g, for which the algorithm diverges.

In this chapter we deal with conditions which ensure the recovery of the exact support I with OMP, and we give an error estimate of the approximate solution. Some analysis in this sense is already available in a sparse approximation setting, cf. e.g. [29, 101]. An important application of sparse approximation is compressive sampling, a possibility of nonadaptive, efficient sampling, utilizing the prior knowledge that the signal is sparse [13, 14, 16].

In sparse approximation one represents or approximates a signal using a small number of waveforms. Since the collection of possible waveforms is typically large, one has to deal with the problem of redundancy of a dictionary, i.e. the atoms φ_i are not linearly independent. However, in sparse approximation one expects the waveforms φ_i to be approximatively orthogonal, assuming that the so-called coherence parameter μ is small. It is defined by

$$\mu := \sup_{i \neq j} \left|\left\langle \tfrac{A\psi_i}{\|A\psi_i\|}, \tfrac{A\psi_j}{\|A\psi_j\|}\right\rangle\right| = \sup_{i \neq j} |\langle \varphi_i, \varphi_j\rangle|.$$

The concept of sparse approximation cannot be transfered to ill-posed inverse problems, since here the atoms φ_i are typically far from orthogonal. The ill-posedness of the operator A typically causes that two atoms φ_i and φ_j, $i \neq j$, look much alike. Therefore, the coherence parameter becomes huge (i.e. $\mu \approx 1$) and the results from sparse approximation theory cannot be transfered to ill-posed inverse problems.

In [101], a sufficient and necessary condition for exact recovery for OMP with exact data g is derived. In [29], it is transfered to noisy signals in terms of the coherence parameter μ. In [31,43], the authors derive another recovery condition for non-disturbed signals g, which works without the concept of coherence. The goal of this chapter is the generalization of the results from [31,43,101] to noisy signals. The proceeding is as follows.

i) In section 4.2 we reflect the conditions for exact recovery with exact data g, which are derived in [101] and [31,43]. We rewrite them in the context of infinite-dimensional inverse problems.

ii) Section 4.3 contains the main theoretical results of chapter 4, namely, the generalization of the results from [31,43,101] to noisy signals. Additionally, we give an error estimate of the approximate solution. It turns out that—under a certain condition—OMP yields an a posteriori regularization method.

iii) In section 4.4 we conclude and forecast on future work.

iv) Later, in chapter 5, we demonstrate the practicability of the deduced recovery conditions in the presence of noise with two examples. In section 5.2 we apply the conditions to an example from mass spectrometry. Here, the data are given as sums of Dirac peaks convolved with a Gaussian kernel. Another example from digital holography is concerned in section 5.3. The data are given as sums of characteristic functions convolved with a Fresnel function. The two examples illustrate that the deduced conditions for exact recovery lead to practically relevant estimates such that one may check a priori, if the experimental setup guarantees exact deconvolution with OMP.

The results have been published in [27,68].

4.2 Exact recovery for exact data

In [101], Tropp gives a sufficient and necessary condition for exact recovery with OMP. Next, we list this result in the language of infinite dimensional inverse problems.

To simplify the statements we introduce the following notations. Define the linear continuous synthesis operator for the dictionary $\Phi = \{\varphi_i\} = \{A\psi_i/\|A\psi_i\|_{\mathcal{H}_2}\}$ by

$$D: \quad \ell^1 \to \mathcal{H}_2, \quad (v_i)_{i\in\mathbb{Z}} \mapsto \sum_{i\in\mathbb{Z}} v_i\varphi_i = \sum_{i\in\mathbb{Z}} v_i \frac{A\psi_i}{\|A\psi_i\|}.$$

Since D is linear and bounded, the Banach space adjoint operator

$$D^* : \mathcal{H}_2 \to (\ell^1)^* = \ell^\infty$$

exists and arises as

$$D^*h = (\langle h, \varphi_i\rangle)_{i\in\mathbb{Z}} = \left(\langle h, \tfrac{A\psi_i}{\|A\psi_i\|}\rangle\right)_{i\in\mathbb{Z}}.$$

Denote the standard basis of ℓ^1 by $\{e_i\}_{i\in\mathbb{Z}}$. Furthermore, for a subset $J \subset \mathbb{Z}$ we denote with $P_J : \ell^1 \to \ell^1$ the projection onto coefficients $j \in J$, $P_J u := \sum_{j\in J} u_j e_j$. With $B^\dagger$ for a finite dimensional linear operator B we term the pseudoinverse operator.

Remark 4.1. Note that the use of ℓ^1 and its dual ℓ^∞ arises naturally in this context. In chapter 3 we have used a synthesis operator for the separable preimage Hilbert space $\mathcal{H}_1$, which maps from ℓ^2. This choice has been necessary for an infinite dimensional separable Hilbert space with an orthonormal basis.

To see this, let $\Theta = \{\vartheta_i\}_{i\in\mathbb{Z}}$ be an orthonormal basis of a Hilbert space $\mathcal{H}$. Then the mapping

$$\mathcal{H} \to \ell^2, \quad h \mapsto (\langle h, \vartheta_i\rangle)_{i\in\mathbb{Z}},$$

is an isomorphism and hence the synthesis operator $D : \ell^2 \to \mathcal{H}$ arises naturally.

In contrast to that, in this chapter we have a unit-normed dictionary which is not orthogonal. Thus, let $\Theta = \{\vartheta_i\}_{i\in\mathbb{Z}}$ be a unit-normed dictionary of a Hilbert space $\mathcal{H}$. Then the synthesis operator

$$D : \ell^1 \to \mathcal{H}, \quad v \mapsto \sum_{i\in\mathbb{Z}} v_i\vartheta_i,$$

is bounded, since for $v \in \ell^1$ it holds that

$$\|Dv\|_{\mathcal{H}} = \Big\|\sum_{i\in\mathbb{Z}} v_i\vartheta_i\Big\|_{\mathcal{H}} \le \sum_{i\in\mathbb{Z}} |v_i|\, \|\vartheta_i\|_{\mathcal{H}} = \|v\|_{\ell^1}.$$

Its Banach space adjoint operator

$$D^* : \mathcal{H} \to \ell^\infty, \quad h \mapsto (\langle h, \vartheta_i \rangle)_{i\in\mathbb{Z}},$$

obviously is bounded, as well, since for $h \in \mathcal{H}$ we get the estimate

$$\|D^* h\|_{\ell^\infty} = \sup_{i\in\mathbb{Z}} |\langle h, \vartheta_i\rangle| \leq \|h\|_{\mathcal{H}}.$$

With these notations we state the following theorem. The proof is contained in [101], however, we include it here for the sake of completeness and due to the conceptual differences between sparse approximation and inverse problems.

Theorem 4.2 (Exact recovery condition for exact data [101]). *Let $u^\diamond \in \ell^0$ with $\operatorname{supp}(u^\diamond) = I$, $f^\diamond = \sum_{i\in\mathbb{Z}} u_i^\diamond \psi_i$ and $g = Af^\diamond$ the measured signal. If the operator $A : \mathcal{B}_1 \to \mathcal{H}_2$ and the dictionary $\Psi = \{\psi_i\}_{i\in\mathbb{Z}}$ fulfill the* exact recovery condition (ERC)

$$\sup_{\varphi\in\Phi(I^\complement)} \|(DP_I)^\dagger \varphi\|_{\ell^1} < 1, \tag{4.3}$$

then OMP with its parameter ε set to 0 recovers $u^\diamond$ exactly.

Proof. OMP chooses the right atom in step $k+1$ if

$$\sup_{i\in I} |\langle r^k, \varphi_i\rangle| > \sup_{i\notin I} |\langle r^k, \varphi_i\rangle|,$$

or, formulated with the synthesis operator D and the projections P_I and $P_{I^\complement}$ onto I and $I^\complement := \mathbb{Z}\setminus I$, respectively, if

$$\frac{\|P_{I^\complement} D^* r^k\|_{\ell^\infty}}{\|P_I D^* r^k\|_{\ell^\infty}} < 1.$$

If OMP has chosen correct atoms in the first k steps, i.e. $I^k \subset I$, then we have $r^k \in \operatorname{span}\Phi(I)$. Moreover, $(P_I D^*)^\dagger P_I D^*$ is the orthogonal projection onto $\operatorname{span}\Phi(I)$ and we get

$$r^k = (P_I D^*)^\dagger P_I D^* r^k.$$

Hence, for a correct choice in step $k+1$ it is sufficient that

$$\frac{\|P_{I^\complement} D^* r^k\|_{\ell^\infty}}{\|P_I D^* r^k\|_{\ell^\infty}} = \frac{\|P_{I^\complement} D^* (P_I D^*)^\dagger P_I D^* r^k\|_{\ell^\infty}}{\|P_I D^* r^k\|_{\ell^\infty}} < 1.$$

Consequently, since

$$\|P_{I^\complement} D^*(P_I D^*)^\dagger P_I D^* r^k\|_{\ell^\infty} \leq \|P_{I^\complement} D^*(P_I D^*)^\dagger\|_{\ell^\infty,\ell^\infty} \|P_I D^* r^k\|_{\ell^\infty}$$

we inductively get the following sufficient conditions for a correct choice for any step

$$\|P_{I^\complement} D^*(P_I D^*)^\dagger\|_{\ell^\infty,\ell^\infty} = \|(DP_I)^\dagger DP_{I^\complement}\|_{\ell^1,\ell^1} < 1. \tag{4.4}$$

Obviously, on the one hand,

$$\sup_{i \in I^\complement} \|(DP_I)^\dagger \varphi_i\|_{\ell^1} \leq \sup_{\substack{\|v\|_{\ell^1}=1 \\ \mathrm{supp}(v)=I^\complement}} \|(DP_I)^\dagger \sum_{i\in\mathbb{Z}} v_i \varphi_i\|_{\ell^1} = \|(DP_I)^\dagger DP_{I^\complement}\|_{\ell^1,\ell^1},$$

and, on the other hand, since $(DP_I)^\dagger$ is linear,

$$\begin{aligned} \sup_{\substack{\|v\|_{\ell^1}=1 \\ \mathrm{supp}(v)=I^\complement}} \|(DP_I)^\dagger \sum_{i\in\mathbb{Z}} v_i \varphi_i\|_{\ell^1} &\leq \sup_{\substack{\|v\|_{\ell^1}=1 \\ \mathrm{supp}(v)=I^\complement}} \sum_{i\in\mathbb{Z}} |v_i| \sup_{\varphi\in\Phi(I^\complement)} \|(DP_I)^\dagger \varphi\|_{\ell^1} \\ &= \sup_{\varphi\in\Phi(I^\complement)} \|(DP_I)^\dagger \varphi\|_{\ell^1}. \end{aligned}$$

This shows that $\|(DP_I)^\dagger DP_{I^\complement}\|_{\ell^1,\ell^1} = \sup_{\varphi\in\Phi(I^\complement)} \|(DP_I)^\dagger \varphi\|_{\ell^1}$ and hence inequality (4.3) is another condition for exact recovery, which is equivalent to condition (4.4). □

Theorem 4.2 gives a sufficient condition for exact recovery with OMP. In [101], Tropp shows that condition (4.3) is even necessary in the sense that if

$$\sup_{\varphi\in\Phi(I^\complement)} \|(DP_I)^\dagger \varphi\|_{\ell^1} \geq 1,$$

then there exists a signal g with support I for which OMP does not recover $u^\diamond$ with $g = Af^\diamond = A\sum u_i^\diamond \psi_i$.

The ERC (4.3) is hard to evaluate. Therefore, Dossal and Mallat [31] and Gribonval and Nielsen [43] derive a weaker sufficient but not necessary recovery condition that depends on inner products of the dictionary atoms of $\Phi(I)$ and $\Phi(I^\complement)$. It can be proved similarly to proposition 3.17. Again, for sake of an easier presentation we define

$$\mathrm{COR}_I := \sup_{i\in I} \sum_{\substack{j\in I \\ j\neq i}} |\langle \varphi_i, \varphi_j\rangle| \quad \text{and} \quad \mathrm{COR}_{I^\complement} := \sup_{i\in I^\complement} \sum_{j\in I} |\langle \varphi_i, \varphi_j\rangle|.$$

Proposition 4.3 (Neumann exact recovery condition [31,43]). *Let $u^\diamond \in \ell^0$ with $\operatorname{supp}(u^\diamond) = I$, $f^\diamond = \sum_{i\in\mathbb{Z}} u_i^\diamond \psi_i$ and $g = Af^\diamond$ the measured signal. If the operator $A : \mathcal{B}_1 \to \mathcal{H}_2$ and the dictionary $\Psi = \{\psi_i\}_{i\in\mathbb{Z}}$ fulfill the* Neumann ERC

$$\mathrm{COR}_I + \mathrm{COR}_{I^\complement} < 1, \tag{4.5}$$

then OMP with its parameter ε set to 0 recovers $u^\diamond$.

Proof. By theorem 4.2, OMP recovers right, if

$$\|(P_I D^* D P_I)^{-1}\|_{\ell^1,\ell^1} \sup_{i\in I^\complement} \|P_I D^* \varphi_i\|_{\ell^1} < 1. \tag{4.6}$$

Split the operator $P_I D^* D P_I$ into diagonal Id and off-diagonal $-B$, i.e. $P_I D^* D P_I =: \mathrm{Id} - B$. The Neumann ERC (4.5) in particular implies that

$$\|B\|_{\ell^1,\ell^1} = \mathrm{COR}_I = \sup_{i\in I} \sum_{\substack{j\in I\\ j\neq i}} |\langle \varphi_i, \varphi_j\rangle| < 1.$$

Then, by using the following Neumann series estimate

$$\|(\mathrm{Id} - B)^{-1}\|_{\ell^1,\ell^1} \leq (1 - \|B\|_{\ell^1,\ell^1})^{-1},$$

we can estimate the first term in (4.6) by

$$\|(P_I D^* D P_I)^{-1}\|_{\ell^1,\ell^1} \leq \frac{1}{1 - \mathrm{COR}_I}.$$

With that and rewriting the second term in (4.6),

$$\sup_{i\in I^\complement} \|P_I D^* \varphi_i\|_{\ell^1} = \sup_{i\in I^\complement} \sum_{j\in I} |\langle \varphi_i, \varphi_j\rangle| = \mathrm{COR}_{I^\complement},$$

we get the statement. □

Remark 4.4. Obviously, the Neumann ERC (4.5) is not necessary for exact recovery. A demonstrative example can be found in $\mathbb{R}^4$ with the signal $g = (1,1,1,0)^\top$ and the unit-normed dictionary $\Phi = \{\varphi_i\}_{i=1}^4$, with $\varphi_1 := (1,0,0,0)^\top$, $\varphi_2 := 2^{-1/2}(1,1,0,0)^\top$, $\varphi_3 := 2^{-1/2}(1,0,1,0)^\top$, $\varphi_4 := (0,0,0,1)^\top$. Here with $I := \{1,2,3\}$ and $I^\complement := \{4\}$ we get

$$|\langle \varphi_1, \varphi_4\rangle| + |\langle \varphi_2, \varphi_4\rangle| + |\langle \varphi_3, \varphi_4\rangle| = 0,$$

but

$$|\langle \varphi_1, \varphi_2\rangle| + |\langle \varphi_1, \varphi_3\rangle| = \sqrt{2} > 1,$$

hence the Neumann ERC is not fulfilled. The ERC (4.3) is nevertheless fulfilled, since in that case $\|(DP_I)^\dagger \varphi_4\|_{\ell^1} = 0$. OMP will then recover exactly, as one could expect by considering that just $\{\varphi_1, \varphi_2, \varphi_3\}$ span the $\mathbb{R}^3$.

This counter-example may be generalized by considering $I \subset \mathbb{Z}$ with linearly independent $\{\varphi_i\}_{i \in I}$, such that

$$\sup_{i \in I^{\complement}} \sum_{j \in I} |\langle \varphi_i, \varphi_j \rangle| = 0 \quad \text{and} \quad \sup_{i \in I} \sum_{\substack{j \in I \\ j \neq i}} |\langle \varphi_i, \varphi_j \rangle| \geq 1.$$

Here the Neumann ERC fails, but for any signal with support I, however, OMP will recover exactly since the atoms φ_i, $i \in I$, and φ_j, $j \in I^{\complement}$, are uncorrelated, and OMP never chooses an atom twice.

Remark 4.5. The sufficient conditions for WOMP with weakness parameter $\omega \in (0, 1]$ are

$$\sup_{\varphi \in \Phi(I^{\complement})} \|(DP_I)^\dagger \varphi\|_{\ell^1} < \omega$$

and

$$\mathrm{COR}_I + \tfrac{1}{\omega}\, \mathrm{COR}_{I^{\complement}} < 1,$$

according to theorem 4.2 and proposition 4.3, respectively. They are proved analogously.

Usually for sparse approximation problems the behavior of redundant dictionaries is characterized as follows.

Definition 4.6. Let $\Theta := \{\vartheta_i\}_{i \in \mathbb{Z}}$ be a dictionary. Then the corresponding *coherence parameter* μ and the *cumulative coherence* $\mu_1(m)$ for a positive integer m are defined as

$$\mu := \sup_{i \neq j} |\langle \vartheta_i, \vartheta_j \rangle|$$

and

$$\mu_1(m) := \sup_{\substack{\Lambda \subset \mathbb{Z} \\ |\Lambda| = m}} \sup_{i \notin \Lambda} \sum_{j \in \Lambda} |\langle \vartheta_i, \vartheta_j \rangle|.$$

Note that $\mu_1(1) = \mu$ and $\mu_1(m) \leq m\mu$ for all $m \in \mathbb{N}$.

Since for the dictionary $\Phi = \{\varphi_i\}_{i\in\mathbb{Z}}$ the inequalities

$$\mathrm{COR}_I \le \mu_1(N-1) \quad \text{and} \quad \mathrm{COR}_{I^\complement} \le \mu_1(N)$$

hold, we get another condition in terms of the cumulative coherence, which is even weaker than the Neumann ERC.

Proposition 4.7 (Exact recovery condition in terms of coherence [101]). *Let $u^\diamond \in \ell^0$ with $\mathrm{supp}(u^\diamond) = I$. Let $f^\diamond = \sum_{i\in\mathbb{Z}} u_i^\diamond \psi_i$ and $g = Af^\diamond$ the measured signal. If the operator $A : \mathcal{B}_1 \to \mathcal{H}_2$ and the dictionary $\Psi = \{\psi_i\}_{i\in\mathbb{Z}}$ lead to the dictionary $\Phi = \{\varphi_i\}_{i\in\mathbb{Z}}$ which fulfills the condition*

$$\mu_1(N-1) + \mu_1(N) < 1, \tag{4.7}$$

then OMP with its parameter ε set to 0 recovers $u^\diamond$.

Remark, that the condition in proposition 4.7 for inverse problems might be unsuitable, since the ill-posed operator typically causes that the coherence parameter μ is close to one. Therefore the cumulative coherence μ_1 can grow large with increasing support.

Remark 4.8. Another major approach for solving sparse approximation problems is the basis pursuit (BP), cf. e.g. [10, 20]. Here one solves the convex optimization problem

$$\min_{u\in\ell^2} \|u\|_{\ell^1} \quad \text{subject to} \quad A\sum u_i\psi_i = g.$$

This idea is closely related to Tikhonov regularization with an ℓ^1 penalty. In [101], it is shown that the ERC (4.3) also ensures the exact recovery by means of BP. Since propositions 4.3 and 4.7 are estimates for the ERC (4.3), and the proofs do not take into account any properties of the OMP algorithm, these results hold here, too.

4.3 Exact recovery in the presence of noise

In [29], Donoho, Elad and Temlyakov transfer Tropp's result [101] to noisy signals. They derive a condition for exact recovery in terms of the coherence parameter μ of a dictionary. This condition is—just as remarked in [29]—an obvious weaker condition. As already mentioned, in particular for ill-posed problems this condition is too restrictive. In the following we give exact recovery conditions in the presence of noise, which are closer to the results of theorem 4.2 and proposition 4.3.

Assume that instead of exact data $g = Af^\diamond \in \mathcal{H}_2$, only a noisy version

$$g^\varepsilon = g + \eta = Af^\diamond + \eta$$

with noise level $\|g - g^\varepsilon\|_{\mathcal{H}_2} = \|\eta\|_{\mathcal{H}_2} \leq \varepsilon$ can be observed. Now, OMP has to stop as soon as the representation error r^k is smaller or equal to the noise level ε, i.e. as soon as $\|r^k\|_{\mathcal{H}_2} \leq \varepsilon$.

Theorem 4.9 (Exact recovery condition in the presence of noise). *Let $u^\diamond \in \ell^0$ with* $\operatorname{supp}(u^\diamond) = I$. *Let $f^\diamond = \sum_{i\in\mathbb{Z}} u_i^\diamond \psi_i$ and $g^\varepsilon = Af^\diamond + \eta$ the noisy data with noise level $\|\eta\|_{\mathcal{H}_2} \leq \varepsilon$ and noise-to-signal ratio*

$$r_{\varepsilon/u} := \frac{\sup_{i\in\mathbb{Z}} |\langle \eta, \varphi_i \rangle|}{\min_{i\in I} |u_i^\diamond| \|A\psi_i\|_{\mathcal{H}_2}}.$$

If the operator $A : \mathcal{B}_1 \to \mathcal{H}_2$ and the dictionary $\Psi = \{\psi_i\}_{i\in\mathbb{Z}}$ fulfill the exact recovery condition in the presence of noise (εERC)

$$\sup_{\varphi\in\Phi(I^\complement)} \|(DP_I)^\dagger \varphi\|_{\ell^1} < 1 - 2\, r_{\varepsilon/u}(1 - \mathrm{COR}_I)^{-1}, \tag{4.8}$$

and $\mathrm{COR}_I < 1$, *then OMP recovers the support I of $u^\diamond$ exactly.*

Proof. We prove the εERC (4.8) analogously to theorem 4.2 by induction. Assume that OMP recovered the correct patterns in the first k steps, i.e.

$$\widehat{f}^k = \sum_{i\in I^k} \widehat{u}_i^k \psi_i,$$

with $I^k \subset I$. Then we get for the residual

$$\begin{aligned} r^k &:= g^\varepsilon - A\widehat{f}^k = g + \eta - A\widehat{f}^k = A\Big(\sum_{i\in I} (u_i^\diamond - \widehat{u}_i^k)\psi_i\Big) + \eta \\ &= \sum_{i\in I} \|A\psi_i\|_{\mathcal{H}_2} (u_i^\diamond - \widehat{u}_i^k)\, \varphi_i + \eta, \end{aligned}$$

hence the noiseless residual $s^k := r^k - \eta = \sum \|A\psi_i\|_{\mathcal{H}_2}(u_i^\diamond - \widehat{u}_i^k)\varphi_i$ has support I. The correlation $|\langle r^k, \varphi_i \rangle|$ for $i \in \mathbb{Z}$ can be estimated from below and above, respectively, via

$$|\langle r^k, \varphi_i \rangle| = |\langle s^k + \eta, \varphi_i \rangle| \gtreqless |\langle s^k, \varphi_i \rangle| \mp |\langle \eta, \varphi_i \rangle|.$$

Hence with

$$\sup_{i\in I^{\complement}} |\langle r^k, \varphi_i\rangle| \le \sup_{i\in I^{\complement}} |\langle s^k, \varphi_i\rangle| + \sup_{i\in\mathbb{Z}} |\langle \eta, \varphi_i\rangle|$$

and

$$\sup_{i\in I} |\langle s^k, \varphi_i\rangle| - \sup_{i\in\mathbb{Z}} |\langle \eta, \varphi_i\rangle| \le \sup_{i\in I} |\langle r^k, \varphi_i\rangle|$$

we get the condition

$$\|P_{I^{\complement}} D^* s^k\|_{\ell^\infty} + \sup_{i\in\mathbb{Z}} |\langle \eta, \varphi_i\rangle| < \|P_I D^* s^k\|_{\ell^\infty} - \sup_{i\in\mathbb{Z}} |\langle \eta, \varphi_i\rangle|,$$

which ensures a right choice in the $(k+1)$-st step. Same as in the proof of theorem 4.2, we can write

$$s^k = (P_I D^*)^\dagger P_I D^* s^k,$$

since $\operatorname{supp}(s^k) \subset I$. With this identity, analogously to theorem 4.2 we get the following sufficient condition for OMP in the presence of noise

$$\sup_{\varphi\in\Phi(I^{\complement})} \|(DP_I)^\dagger \varphi\|_{\ell^1} < 1 - 2\,\frac{\sup_{i\in\mathbb{Z}} |\langle \eta, \varphi_i\rangle|}{\|P_I D^* s^k\|_{\ell^\infty}}.$$

The last thing we have to afford to finish the proof is an estimation for the term $\|P_I D^* s^k\|_{\ell^\infty}$ from below.

In the first step this is easy, since $r^0 = g^\varepsilon$, hence $s^0 = g$ with $g = Af^\diamond = A\sum_{i\in I} u_i^\diamond \psi_i$. With that, we get for all $l \in I$

$$\begin{aligned}
\|P_I D^* s^0\|_{\ell^\infty} &= \|P_I D^* g\|_{\ell^\infty} = \sup_{j\in I} |\langle g, \varphi_j\rangle| \\
&= \sup_{j\in I} \Big| \sum_{i\in I} u_i^\diamond \|A\psi_i\|_{\mathcal{H}_2} \langle \varphi_i, \varphi_j\rangle \Big| \ge \Big| \sum_{i\in I} u_i^\diamond \|A\psi_i\|_{\mathcal{H}_2} \langle \varphi_i, \varphi_l\rangle \Big| \\
&= \Big| u_l^\diamond \|A\psi_l\|_{\mathcal{H}_2} + \sum_{\substack{i\in I\\ i\ne l}} u_i^\diamond \|A\psi_i\|_{\mathcal{H}_2} \langle \varphi_i, \varphi_l\rangle \Big| \\
&\ge |u_l^\diamond| \|A\psi_l\|_{\mathcal{H}_2} - \sum_{\substack{i\in I\\ i\ne l}} |u_i^\diamond| \|A\psi_i\|_{\mathcal{H}_2} |\langle \varphi_i, \varphi_l\rangle|.
\end{aligned}$$

Choose $m \in \arg\max_{l\in I}\{|u_l^\diamond| \|A\psi_l\|_{\mathcal{H}_2}\}$. Then

$$\begin{aligned}
\|P_I D^* s^0\|_{\ell^\infty} &\ge |u_m^\diamond| \|A\psi_m\|_{\mathcal{H}_2} - |u_m^\diamond| \|A\psi_m\|_{\mathcal{H}_2} \sum_{\substack{i\in I\\ i\ne m}} |\langle \varphi_i, \varphi_m\rangle| \\
&\ge |u_m^\diamond| \|A\psi_m\|_{\mathcal{H}_2} (1 - \mathrm{COR}_I\,).
\end{aligned}$$

By assumption $\mathrm{COR}_I < 1$, and obviously it holds that $|u_m^\diamond| \|A\psi_m\|_{\mathcal{H}_2} \geq \min_{i\in I} |u_i^\diamond| \, \|A\psi_i\|_{\mathcal{H}_2}$, hence we get

$$\|P_I D^* s^0\|_{\ell^\infty} \geq \min_{i\in I} |u_i^\diamond| \, \|A\psi_i\|_{\mathcal{H}_2} \big(1 - \mathrm{COR}_I\big).$$

To prove this for general k we successively apply this estimation. Again, we get for all $l \in I$

$$\begin{aligned}
\|P_I D^* s^k\|_\infty &= \sup_{j\in I} |\langle s^k, \varphi_j\rangle| = \sup_{j\in I} \Big| \sum_{i\in I} (u_i^\diamond - \widehat{u}_i^k) \|A\psi_i\|_{\mathcal{H}_2} \langle \varphi_i, \varphi_j\rangle \Big| \\
&\geq \Big| \sum_{i\in I} (u_i^\diamond - \widehat{u}_i^k) \|A\psi_i\|_{\mathcal{H}_2} \langle \varphi_i, \varphi_l\rangle \Big| \\
&\geq |u_l^\diamond - \widehat{u}_l^k| \|A\psi_l\|_{\mathcal{H}_2} - \sum_{\substack{i\in I \\ i\neq l}} |u_i^\diamond - \widehat{u}_i^k| \|A\psi_i\|_{\mathcal{H}_2} |\langle \varphi_i, \varphi_l\rangle|.
\end{aligned}$$

Choose $m \in \arg\max_{l\in I} \{ |u_l^\diamond - \widehat{u}_l^k| \|A\psi_l\|_{\mathcal{H}_2} \}$. Then

$$\|P_I D^* s^0\|_{\ell^\infty} \geq |u_m^\diamond - \widehat{u}_m^k| \|A\psi_m\|_{\mathcal{H}_2} \big(1 - \mathrm{COR}_I\big).$$

Since

$$|u_m^\diamond - \widehat{u}_m^k| \|A\psi_m\|_{\mathcal{H}_2} \geq \max_{l\in I, l\notin I^k} |u_l^\diamond - \widehat{u}_l^k| \|A\psi_l\|_{\mathcal{H}_2} = \max_{l\in I, l\notin I^k} |u_l^\diamond| \|A\psi_l\|_{\mathcal{H}_2}$$

we get

$$\|P_I D^* s^k\|_{\ell^\infty} \geq \min_{i\in I} |u_i^\diamond| \, \|A\psi_i\|_{\mathcal{H}_2} \big(1 - \mathrm{COR}_I\big).$$

□

To ensure the εERC (4.8) one has necessarily for the noise-to-signal ratio $r_{\varepsilon/u} < 1/2$. For a small noise-to-signal ratio the εERC (4.8) approximates the ERC (4.3). A rough upper bound for $\sup_{i\in\mathbb{Z}} |\langle \eta, \varphi_i\rangle|$ is ε and hence, one may use

$$r_{\varepsilon/u} \leq \frac{\varepsilon}{\min\limits_{i\in I} |u_i^\diamond| \|A\psi_i\|_{\mathcal{H}_2}}.$$

Similar to the noiseless case, the εERC (4.8) is hard to evaluate. Analogously to section 4.2, we now give a weaker sufficient recovery condition that depends on inner products of the dictionary atoms. It can be proved analogously to proposition 4.3.

Proposition 4.10 (Neumann ERC in the presence of noise). *Let* $u^\diamond \in \ell^0$ *with* $\operatorname{supp}(u^\diamond) = I$. *Let* $f^\diamond = \sum_{i\in\mathbb{Z}} u_i^\diamond \psi_i$ *and* $g^\varepsilon = Af^\diamond + \eta$ *the noisy data with noise level* $\|\eta\|_{\mathcal{H}_2} \le \varepsilon$ *and noise-to-signal ratio* $r_{\varepsilon/u}$. *If the operator* $A : \mathcal{B}_1 \to \mathcal{H}_2$ *and the dictionary* $\Psi = \{\psi_i\}_{i\in\mathbb{Z}}$ *fulfill the* Neumann εERC

$$\mathrm{COR}_I + \mathrm{COR}_{I^\complement} < 1 - 2\, r_{\varepsilon/u}, \tag{4.9}$$

then OMP recovers the support I *of* $u^\diamond$ *exactly.*

Remark 4.11. The sufficient conditions for WOMP for the case of noisy data with weakness parameter $\omega \in (0, 1]$ are

$$\sup_{\varphi\in\Phi(I^\complement)} \|(DP_I)^\dagger \varphi\|_{\ell^1} < \omega - 2\, r_{\varepsilon/u}(1 - \mathrm{COR}_I)^{-1},$$

and

$$\mathrm{COR}_I + \tfrac{1}{\omega}\, \mathrm{COR}_{I^\complement} < 1 - \tfrac{1}{\omega}\, 2\, r_{\varepsilon/u},$$

according to theorem 4.9 and proposition 4.10, respectively.

Remark 4.12. If $A\psi_i$ is unit-normed for all $i \in \mathbb{Z}$, then the Neumann εERC for OMP (4.9) and the Neumann εERC for the ℓ^1-penalized Tikhonov functional (3.26) in proposition 3.17 are equal.

Same as for the noiseless case, we can give another even weaker condition in terms of the cumulative coherence of the dictionary Φ.

Proposition 4.13 (Exact recovery condition in terms of coherence). *Let* $u^\diamond \in \ell^0$ *with* $\operatorname{supp}(u^\diamond) = I$. *Let* $f^\diamond = \sum_{i\in\mathbb{Z}} u_i^\diamond \psi_i$ *and* $g^\varepsilon = Af^\diamond + \eta$ *the noisy data with noise level* $\|\eta\|_{\mathcal{H}_2} \le \varepsilon$ *and noise-to-signal ratio* $r_{\varepsilon/u}$. *If the operator* $A : \mathcal{B}_1 \to \mathcal{H}_2$ *and the dictionary* $\Psi = \{\psi_i\}_{i\in\mathbb{Z}}$ *lead to the dictionary* $\Phi = \{\varphi_i\}_{i\in\mathbb{Z}}$ *which fulfills the condition*

$$\mu_1(N-1) + \mu_1(N) < 1 - 2\, r_{\varepsilon/u}, \tag{4.10}$$

then OMP recovers the support I *of* $u^\diamond$ *exactly.*

Theorem 4.9 and proposition 4.10 just ensure the correct support I. The following simple proposition shows that the reconstruction error is of the order of the noise level.

Proposition 4.14 (Error bounds in the presence of noise). *Let the assumptions of theorem 4.9 be fulfilled. Then there exists a constant* $C > 0$, *such that for the approximative solution $\widehat{u}$ determined by OMP the following error bound holds:*

$$\|\widehat{u} - u^\diamond\|_{\ell^1} \leq C\varepsilon.$$

Hence in this case OMP yields an a posteriori regularization method with a linear convergence rate.

Proof. Since the εERC (4.8) is fulfilled, OMP recovered the correct support I, i.e.

$$\widehat{u} = \arg\min \left\{ \|g^\varepsilon - \textstyle\sum_{i\in I} \breve{u}_i A\psi_i\|_{\mathcal{H}_2} \,\middle|\, \breve{u} \in \ell^2(I) \right\}.$$

With the help of the operator $K_I : \ell^1(I) \to \mathcal{H}_2$ defined by $K_I v = \sum_{i\in I} v_i A\psi_i$, this is equivalently written as

$$K_I^* K_I \widehat{u} = K_I^* g^\varepsilon.$$

Note that $K_I^* K_I : \ell^2(I) \to \ell^2(I)$ is just the matrix

$$(K_I^* K_I)_{i,j} = \langle A\psi_i, A\psi_j \rangle.$$

For the error we get

$$\|\widehat{u} - u^\diamond\|_{\ell^1} = \|K_I^\dagger (g^\varepsilon - g)\|_{\ell^1} \leq \|K_I^\dagger\|_{\mathcal{H}_2,\ell^1} \, \|(g^\varepsilon - g)\|_{\mathcal{H}_2} = C\varepsilon.$$

□

Remark 4.15. We remark again on an exact recovery conditions for BP. For noisy data g^ε, the BP minimization problem reads

$$\min_{u\in\ell^2} \|u\|_{\ell^1} \quad \text{subject to} \quad \|A \textstyle\sum u_i\psi_i - g^\varepsilon\|_{\mathcal{H}_2} \leq \varepsilon.$$

Unlike in section 4.2 where the results can be transfered to BP, see remark 4.8, this is not possible for the presence of noise. To prove theorem 4.9, we used properties of the OMP algorithm, which are not valid for BP.

For the case of noisy data g^ε in [29] an exact recovery condition for BP is derived. This condition depends on the coherence parameter μ of the dictionary Φ. In this chapter the focus is on the *greedy* solution of inverse problems, hence we give up on deriving stronger results for BP, which are closer to the results here.

Remark 4.16 (Nonnegativity constraints)**.** Sometimes in inverse problems a nonnegativity constraint is valid, i.e. one knows a priori for the source $f^\diamond$,

$$f^\diamond = \sum_{i\in\mathbb{Z}} u_i^\diamond \psi_i,$$

that the coefficients $u_i^\diamond$ are nonnegative, i.e. $u_i^\diamond \geq 0$ for all $i \in \mathbb{Z}$. This is e.g. the case in the examples from mass spectrometry and digital holography of particles discussed in chapter 5.

In [11], a variant of OMP it presented, which takes into account a nonnegativity constraint via modifying the greedy selection step to

$$i_k \in \arg\sup \left\{ \langle r^{k-1}, \varphi_i \rangle \,\middle|\, i \in \mathbb{Z} \right\},$$

i.e. without the modulus $|\cdot|$. Analogously, the projection step is changed to

$$\widehat{f}^k := \arg\min \left\{ \|g^\varepsilon - A\widehat{f}\|^2_{\mathcal{H}_2} \,\middle|\, \widehat{f} = \textstyle\sum_i \widehat{u}_i \psi_i,\ \widehat{u}_i \geq 0,\ \mathrm{supp}(\widehat{u}) \subset I^k \right\},$$

compare with algorithm 4.1, page 70.

In [11], the authors remark that the results for exact recovery of OMP with exact data g as developed in [101] also hold for this modified algorithm. To expatiate on that, since $\langle r^k, \varphi_i \rangle \geq 0$, $i \in I$, it is sufficient for a right choice in step $k+1$ if

$$\sup_{i\in I} |\langle r^k, \varphi_i \rangle| > \sup_{i\notin I} |\langle r^k, \varphi_i \rangle|,$$

since

$$\sup_{i\in I} \langle r^k, \varphi_i \rangle = \sup_{i\in I} |\langle r^k, \varphi_i \rangle| > \sup_{i\notin I} |\langle r^k, \varphi_i \rangle| > \sup_{i\notin I} \langle r^k, \varphi_i \rangle.$$

The remain of the proof goes identically to the proof of theorem 4.2.

The results for a noisy observation g^ε as presented in this section are valid, as well. This can be seen by splitting the noisy residual r^k into noise η and noiseless residual s^k with support I, see proof of theorem 4.9. With that, the same justification as above for s^k leads to the proposed results.

These identical recovery conditions, however, are dissatisfactory, since the additional prior knowledge, i.e. $u^\diamond \geq 0$, is not utilized. An idea to come to stronger recovery conditions with nonnegativity constraints is to abstain from the estimation $|\langle r^k, \varphi_i \rangle| > \langle r^k, \varphi_i \rangle$ for $i \notin I$. Instead of that one could work with the following sufficient condition for recovery in step $k+1$

$$\frac{\sup\{P_{I^\complement} D^* r^k\}}{\sup\{P_I D^* r^k\}} < 1.$$

We postpone this idea for future work.

4.4 Conclusion

In this chapter we have given exact recovery conditions for the orthogonal matching pursuit for noisy signals, that work without the coherence parameter. Our motivation has been to treat ill-posed problems. We have obtained results on exact recovery of the support for noisy data. Moreover, there is a simple error bound in which shows a convergence rate of $\mathcal{O}(\varepsilon)$. The rate of convergence resembles what is known for sparsity-enforcing regularization with an ℓ^p-penalty term, with $0 < p \leq 1$ [9, 40, 41].

Granted, to apply the Neumann εERC (4.9) one has to know the support I. However, once there is a sufficiently decaying upper bound for the correlations

$$\mathrm{COR}_I := \sup_{i \in I} \sum_{\substack{j \in I \\ j \neq i}} |\langle \varphi_i, \varphi_j \rangle| \quad \text{and} \quad \mathrm{COR}_{I^\complement} := \sup_{i \in I^\complement} \sum_{j \in I} |\langle \varphi_i, \varphi_j \rangle|,$$

we are able to apply the Neumann εERC (4.9) and obtain a priori computable conditions for exact recovery. In chapter 5 we illustrate the practical relevance with two examples, namely, an example from mass spectrometry (section 5.2) and an example from digital holography of particles (section 5.3).

An idea to come to a tighter exact recovery condition is to bring in more prior knowledge, as e.g. a nonnegativity constraint, cf. remark 4.16.

In a lot of applications, hoping for definite exact recovery is too optimistic. Often it is not possible to obtain exact recovery for all signals but for almost all, i.e. with a high propability. Hence it would be useful to have probabilistic results for exact reconstruction with OMP. In [2], it is shown that OMP exhibits a so-called *phase transition property* if A is a Gaussian matrix. That is, the algorithm either finds the correct support with a high probability, or with a high probability the support cannot be found. It is interesting, if it is possible to obtain phase transitions for some special ill-posed operators, as well. We postpone these ideas for future work.

A straightforward generalization of our approach to fully continuous dictionaries $\Phi = \{\varphi_t\}_{t\in\mathbb{R}}$ runs into problems. Especially, it seems that there is little hope to obtain exact recovery of the support, but maybe one may obtain bounds on how accurate the support is localized. This is strongly related to the structure of the dictionary and of course related to the correlations. We go into this detail in section 5.2 later on.

So far, we proved regularization results just if the exact recovery condition is fulfilled. Furthermore, it would be intersting to give conditions on the operator A and the dictionary Ψ, that ensure OMP to be a regularization method, without postulating that OMP recovers the correct support. I.e., on which condition on A and Ψ we have

$$\hat{f}^{\varepsilon} \to f^{\diamond}, \quad \varepsilon \to 0,$$

in an appropriate norm or distance. Here $\hat{f}^{\varepsilon}$ denotes the approximate solution determined by the orthogonal matching pursuit with data g^{ε}.

Finally, a further direction of research may be to investigate other types of iterative greedy algorithms, as e.g. the Stagewise Orthogonal Matching Pursuit (StOMP) [30], the Regularized Orthogonal Matching Pursuit (ROMP) [80–82], the Compressive Sampling Matching Pursuit (CoSaMP) [79] or the Subspace Pursuit (SP) [24].

5 Application of exact recovery conditions

5.1 Introduction

To apply the Neumann εERCs (3.26) and (4.9), one has to know the support I. In this case, there is no need to apply complex reconstruction methods. One may just solve the restricted least squares problem. For deconvolution problems, however, with a certain prior knowledge, it is possible to evaluate the Neumann εERCs (3.26) and (4.9) a priori, especially when the support I is not known exactly.

In the following we use the Neumann εERCs (3.26) and (4.9) exemplarily with impulse trains convolved with a Gaussian kernel, as e.g. occurs in mass spectrometry, cf. [6, 23, 57], and for characteristic functions convolved with a Fresnel function, as e.g. used in digital holography of particles, cf. [26, 38, 97]. For the sake of an easier presentation, we normalize the operator A such that $\|A\psi_i\|_{\mathcal{H}_2} = 1$ for all $i \in \mathbb{Z}$. As remarked, the Neumann conditions for the ℓ^1-penalized Tikhonov functional (3.26) and for OMP (4.9) are equal in this case.

Remark 5.1. As mentioned in chapter 4, the orthogonal matching pursuit was first proposed in the signal processing context in [73] and [83]. However, for deconvolution problems in radio astronomy the algorithm is well-known for a long time under the name of CLEAN, proposed by Högbom in 1974 [50].

The results of this chapter have been published in [27].

5.2 Mass spectrometry

5.2.1 Introduction

Before we start with the inverse problem model, we give a brief introduction to mass spectrometry, which is based on [75, 107].

Mass spectrometry is a method of analytical chemistry to determine the elemental composition of a chemical sample. This task is accomplished through the experimental measurement of the mass of gas-phase ions produced from molecules of the sample to be examined. Mass spectrometry does not directly determine mass, but it determines the mass-to-charge ratio (m/z) of the ions.

In the following we explain the so-called matrix-assisted laser desorption/ionization time-of-flight (MALDI-TOF) mass spectrometer as the representative for mass spectrometry techniques, cf. figure 10. In MALDI mass spectrometry the sample molecules are embedded into a matrix together with crystallized molecules. The ionization of the sample molecules is triggered by a laser beam. After the laser desorption, the ions are accelerated by an electric field. Since the velocity depends on the mass-to-charge ratio, it is possible to find the mass-to-charge ratio of the ions by measuring their time-of-flight (TOF).

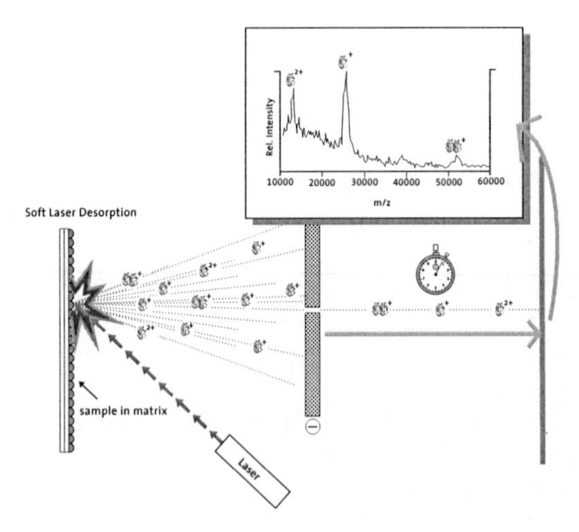

Figure 10: Schematic presentation of MALDI-TOF, from [75]. (Illustration: Typoform Copyright: The Royal Swedish Academy of Sciences.)

Mass spectrometry plays an important role in biology and medice, e.g. for the characterization of proteins in drug discovery or for analyzing urine and blood of people with diseases. However, mass spectrometry is not limited to analysis of organic molecules. It can be used for the detection of any element that can be ionized, and it is e.g. used to analyze silicon wafers to determine the presence of lead and iron.

The data representation in mass spectrometry are the mass spectra. In figure 11 an example of a mass spectrum is displayed. Here, the mass-to-charge ratio m/z corresponds to the x-axis and the intensity corresponds to the y-axis. It is possible to include additional experimental parameters in additional dimensions, such as a retention time according to physio-chemical properties or a spatial coordinate, as in liquid chromatography mass spectrometry (LC/MS) [62] or MALDI imaging [17], respectively.

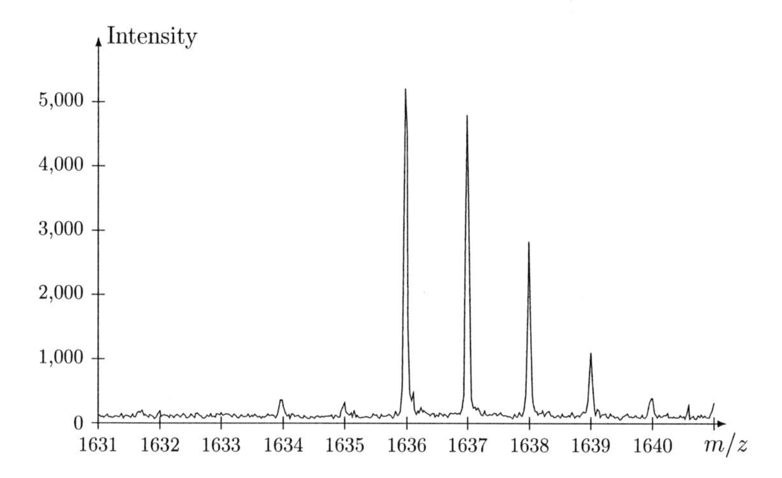

Figure 11: Example of a mass spectrum.

5.2.2 Data model

In mass spectrometry the source $f^\diamond$ is given—after simplification—as sum of Dirac peaks at integer positions $i \in \mathbb{Z}$,

$$f^\diamond = \sum_{i \in \mathbb{Z}} u_i^\diamond \, \delta(\cdot - i),$$

with $|\operatorname{supp}(u^\diamond)| = |I| = N$. Since the measuring procedure is influenced by Gaussian noise, the measured data can be modeled by a convolution operator A with Gaussian kernel

$$\kappa(x) = \frac{1}{\pi^{1/4}\sigma^{1/2}} \exp\Big(-\frac{x^2}{2\sigma^2}\Big), \tag{5.1}$$

i.e. in mass spectrometry the operator equation under consideration is given by, cf. e.g. [6, 23, 57],

$$Af^\diamond = \kappa * f^\diamond = g.$$

Hence, the dictionary which spans the space of all possible signals g reads

$$\Phi := \{\varphi_i\}_{i\in\mathbb{Z}} := \{\kappa * \delta(\cdot - i)\}_{i\in\mathbb{Z}} = \{\kappa(\cdot - i)\}_{i\in\mathbb{Z}}.$$

Notice, that the normalization in (5.1) implies that $\|\kappa(\cdot - i)\|_{L^2} = 1$ for all $i \in \mathbb{Z}$.

As preimage space $\mathcal{B}_1$ we may use the space $\mathcal{M}$ of regular Borel measures on $\mathbb{R}$ (which contains impulse trains if the coefficients $u_i^\diamond$ are summable). If $\mathcal{B}_1$ should be a Hilbert space $\mathcal{H}_1$, then we may use the Sobolev space $H^{-\frac{1}{2}-\varsigma}(\mathbb{R})$, with $\varsigma > 0$. The choice of the image space $\mathcal{H}_2$ is more obvious. Here we use the Lebesgue space $L^2(\mathbb{R})$.

5.2.3 Resolution bounds for mass spectrometry

To verify the εERCs (3.26) and (4.9), we need the autocorrelation of two atoms $\kappa(\cdot - i)$ and $\kappa(\cdot - j)$. In $L^2(\mathbb{R})$ it arises as

$$\begin{aligned}\langle\kappa(\cdot - i), \kappa(\cdot - j)\rangle_{L^2} &= \int_{\mathbb{R}} \tfrac{1}{\sqrt{\pi}\sigma} \exp\big(-\tfrac{(x-i)^2}{2\sigma^2}\big)\exp\big(-\tfrac{(x-j)^2}{2\sigma^2}\big)\,\mathrm{d}\,x \\ &= \exp\big(-\tfrac{(i-j)^2}{4\sigma^2}\big),\end{aligned} \tag{5.2}$$

which is positive and monotonically decreasing in the distance $|i - j|$. If we additionally assume that the peaks of $f^\diamond$ have the minimal distance

$$\rho := \min_{i,j\in\operatorname{supp}(u^\diamond)} |i - j|,$$

then we can estimate the sums of correlations COR_I and $\mathrm{COR}_{I^\complement}$ from above as follows, compare figure 12. W.l.o.g. we fix one peak at $i = 0$ and estimate with the worst case that the peaks appear at position $j\rho$

with $-\lfloor N/2 \rfloor \leq j \leq \lfloor N/2 \rfloor$. Then, for $\rho \in \mathbb{N}$ we get for the correlations of support atoms

$$\mathrm{COR}_I = \sup_{i \in I} \sum_{\substack{j \in I \\ j \neq i}} |\langle \varphi_i, \varphi_j \rangle| \leq 2 \sum_{j=1}^{\lfloor N/2 \rfloor} \langle \kappa, \kappa(\cdot - j\rho) \rangle = 2 \sum_{j=1}^{\lfloor N/2 \rfloor} \exp\big(-\tfrac{(j\rho)^2}{4\sigma^2}\big).$$

For the correlations of support atoms and non-support atoms we have to distinguish between two cases for ρ. For $\rho \geq 2$ we get

$$\begin{aligned}\mathrm{COR}_{I^\complement} = \sup_{i \in I^\complement} \sum_{j \in I} |\langle \varphi_i, \varphi_j \rangle| &\leq \sup_{1 \leq i < \rho} \sum_{j=-\lfloor N/2 \rfloor}^{\lfloor N/2 \rfloor} \langle \kappa(\cdot - i), \kappa(\cdot - j\rho) \rangle \\ &= \sup_{1 \leq i < \rho} \sum_{j=-\lfloor N/2 \rfloor}^{\lfloor N/2 \rfloor} \exp\big(-\tfrac{(i-j\rho)^2}{4\sigma^2}\big)\end{aligned}$$

and for $\rho = 1$

$$\begin{aligned}\mathrm{COR}_{I^\complement} = \sup_{i \in I^\complement} \sum_{j \in I} |\langle \varphi_i, \varphi_j \rangle| &\leq 2 \sum_{j=1}^{\lfloor N/2 \rfloor + 1} \langle \kappa, \kappa(\cdot - j) \rangle \\ &= 2 \sum_{j=1}^{\lfloor N/2 \rfloor + 1} \exp\big(-\tfrac{j^2}{4\sigma^2}\big).\end{aligned}$$

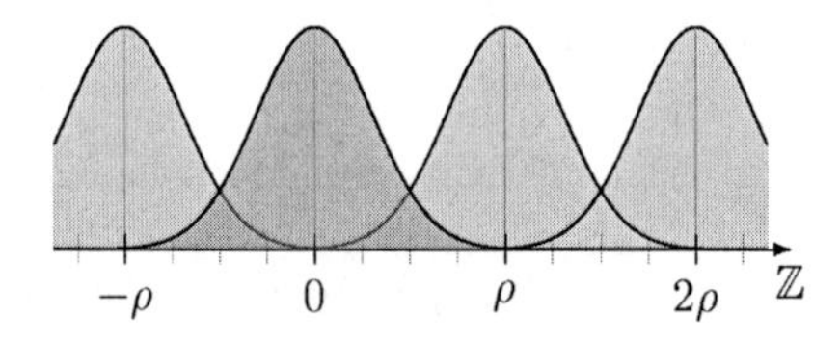

Figure 12: Assume that the peaks have the minimal distance ρ.

Remark 5.2. If the cardinality N of the support I is unknown, then one can replace the finite sums by infinite sums. Obviously, these sums exist since the geometric series is a majorizing series. With ι representing the imaginary unit, they can be expressed in terms of the Jacobi theta function of the third kind, $\vartheta_3(z,q) := \sum_{j=-\infty}^{\infty} q^{j^2} \exp(2j\iota z)$.

With that, we can formulate an estimation for the Neumann εERCs for the ℓ^1-penalized Tikhonov regularization (3.26) and OMP (4.9).

Proposition 5.3 (Neumann εERC for Gaussian-convolved Dirac peaks). *An estimation from above for the Neumann εERCs (3.26) and (4.9) for Dirac peaks convolved with a Gaussian kernel is for $\rho \geq 2$*

$$2 \sum_{j=1}^{\lfloor N/2 \rfloor} \exp\left(-\tfrac{(j\rho)^2}{4\sigma^2}\right) + \sup_{1 \leq i < \rho} \sum_{j=-\lfloor N/2 \rfloor}^{\lfloor N/2 \rfloor} \exp\left(-\tfrac{(i-j\rho)^2}{4\sigma^2}\right) < 1 - 2r_{\varepsilon/u}, \quad (5.3)$$

and for $\rho = 1$

$$2 \sum_{j=1}^{\lfloor N/2 \rfloor} \exp\left(-\tfrac{j^2}{4\sigma^2}\right) + 2 \sum_{j=1}^{\lfloor N/2 \rfloor + 1} \exp\left(-\tfrac{j^2}{4\sigma^2}\right) < 1 - 2r_{\varepsilon/u}. \quad (5.4)$$

This means, that there is a regularization parameter α which allows exact recovery of the support with the ℓ^1-penalized Tikhonov regularization, and that we are able to recover the support with OMP exactly, if the above conditions are fulfilled.

Remark 5.4. The case $\rho = 1$ of the proposition 5.3 coincides with the recovery condition in terms of the cumulative coherence, see remark 3.19 and proposition 4.13. For an odd N we get

$$\mu_1(N) = 2 \sum_{j=1}^{\lfloor N/2 \rfloor} \exp(-j^2/(4\sigma^2)).$$

For the conditions in terms of the cumulative coherence, summing up just over a subset of $\rho\mathbb{Z} := \{j \in \mathbb{Z} \,|\, j/\rho \in \mathbb{Z}\}$ is not a feasible estimation, since for the support I we allow any point $i \in \mathbb{Z}$ and not only atoms of the sub-dictionary $\Phi(\rho\mathbb{Z})$. This turns out to be the main disadvantage of the coherence conditions. They do not distinguish between support and non-support atoms, and hence they give weaker estimations.

In figure 13 conditions (5.3) and (5.4) of proposition 5.3 are plotted for some combinations of σ, ρ and $r_{\varepsilon/u}$ with unknown N. The colored areas describe the combinations where the conditions are fulfilled.

Often for deconvolution problems, the autocorrelation of two atoms $|\langle \varphi(\cdot - i), \varphi(\cdot - j)\rangle|$ is not monotonically decreasing in the distance $|i - j|$, and it obviously depends on the kernel κ. However, if the correlation of

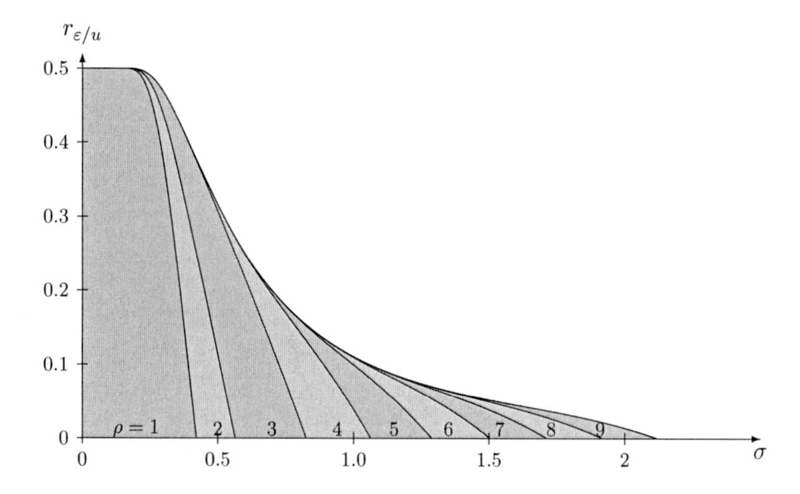

Figure 13: The Neumann εERCs (5.3) and (5.4) for combinations of σ, ρ and $r_{\varepsilon/u}$.

two atoms can be estimated from above via a monotonically decreasing function with respect to an appropriate distance, then we can use a similar estimate. We do this exemplarily for an oscillating kernel in section 5.3, namely, for Fresnel-convolved characteristic functions as appear in digital holography.

Remark 5.5. We remark on a possible fully continuous formulation of OMP. We assume that we are given some data

$$g = \kappa * f^\diamond = \sum_{i \in \mathbb{Z}} u_i^\diamond \, \kappa(\cdot - x_i),$$

and that we do not know the positions x_i. We allow our dictionary to be uncountable, i.e. we search for peaks at every real number. Note that here $i \in \mathbb{Z}$ does not represent the set of possible positions for peaks, but it is an index set for continuous positions $x_i \in \mathbb{R}$.

In the first step of the orthogonal matching pursuit we correlate g with $\kappa(\cdot - x)$ and take that x which gives maximal correlation. In the special case of the Gausssian blurring kernel (5.1) this amounts in finding the maximum of the function

$$F(x) = |\langle g, \kappa(\cdot - x)\rangle| = \sum_{i\in\mathbb{Z}} u_i^\diamond \langle \kappa(\cdot - x_i), \kappa(\cdot - x)\rangle.$$

From (5.2) we see that this is

$$F(x) = \sum_{i\in\mathbb{Z}} u_i^\diamond \exp\Big(-\frac{(x-x_i)^2}{4\sigma^2}\Big).$$

It is clear, that any maximum of F is unlikely to be precisely at some of the x_i's, albeit very close. A detailed study of this effect goes beyond the scope of this chapter, and we present a simple example.

Let us assume that we have two peaks, one at 0 and one at x_1, i.e.

$$f^\diamond = u_0^\diamond \delta(\cdot) + u_1^\diamond \delta(\cdot - x_1). \tag{5.5}$$

Moreover, we assume that $u_0^\diamond > u_1^\diamond$, i.e. the peak in zero is higher. The orthogonal matching pursuit find the first peak at the maximum of the function

$$F(x) = u_0^\diamond \exp\Big(-\frac{x^2}{4\sigma^2}\Big) + u_1^\diamond \exp\Big(-\frac{(x-x_1)^2}{4\sigma^2}\Big),$$

and hence at some root of

$$F'(x) = -\tfrac{1}{2\sigma^2}\Big(u_0^\diamond x \exp\big(-\tfrac{x^2}{4\sigma^2}\big) + u_1^\diamond (x-x_1)\exp\big(-\tfrac{(x-x_1)^2}{4\sigma^2}\big)\Big).$$

The error that OMP makes is hence the error ϱ in the root of F' near zero. In figure 14 it is shown, how the root of F' close to zero depends on the variance σ. One observes that the error ϱ is smaller than the variance σ by some orders of magnitude.

As a final remark we mention that we measured the error not in some norm but only the distance of the δ-peaks. This corresponds to the so-called *Prokhorov metric* which is a metrization for the weak-* convergence in measure space. Convergence in the Prokhorov metric for regularization methods for ill-posed problems has for example been studied in [33].

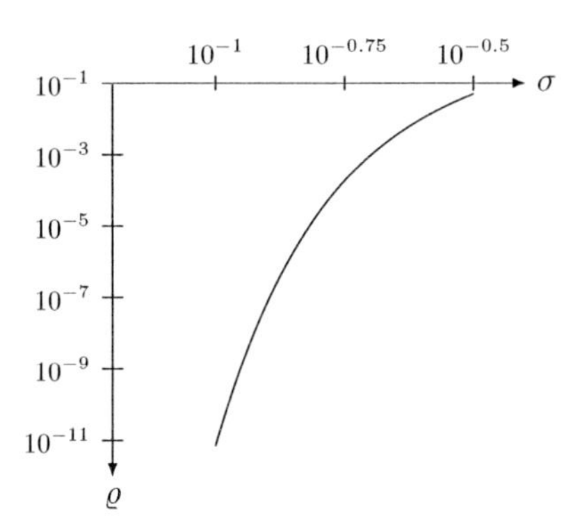

Figure 14: Error of the first step of the matching pursuit for the signal (5.5) with $u_0^\diamond = 2$, $u_1^\diamond = 1$ and $x_1 = 1$. The variable σ is the variance of the Gaussian kernel and ϱ is the position at which the orthogonal matching pursuit locates the first peak.

5.2.4 Numerical examples

We apply the Neumann εERC of proposition 5.3, i.e. conditions (5.3) and (5.4), to simulated data of an isotope pattern. Here the data consist of equidistant peaks with different heights. In our example we use four peaks with a distance of $\rho = 5$ and heights of 3500, 4000, 3800 and 3000. In figure 15 this isotope pattern is represented by the dots. After convolving with Gaussian kernel with $\sigma = 1.125$ we apply a Poisson noise model. This noise model is realistic, because in mass spectrometry a finite number of particles is counted.

In the first example with low noise the Neumann εERC from proposition 5.3 is fulfilled, and hence the support is recovered exactly, both for the ℓ^1-penalized Tikhonov regularization and for OMP, see top of figure 15. However, the condition is restrictive. For the second example the signal is disturbed with huge noise and the Neumann εERC from proposition 5.3 is not fulfilled, see bottom of figure 15. Certainly, both the ℓ^1-penalized Tikhonov regularization and OMP recovered the support exactly. For minimization of the Tikhonov functional the iterated soft-thresholding algorithm [25] was used.

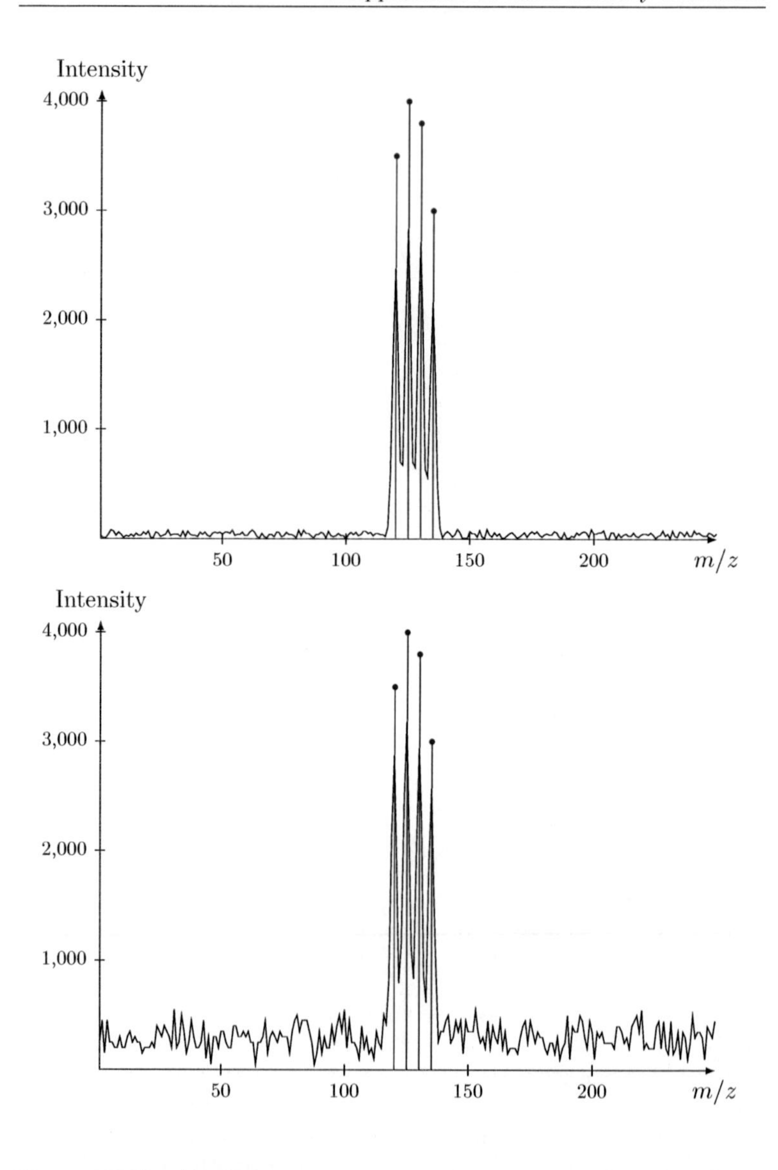

Figure 15: Simulated isotope patterns, represented by the dots. Top: Gaussian-convolved data with low noise, Neumann εERC is satisfied. Bottom: Gaussian-convolved data with high noise, Neumann εERC is not satisfied, but it is still exact recovery possible.

5.3 Digital holography

5.3.1 Introduction

Before we start with the data model, we give a brief introduction to holography and digital holography. This paragraph is based on [94].

Holography is a technique invented by Dennis Gabor in 1948 for recording and reconstructing amplitude and phase of a wave field. The word holography is composed of the Greek words "holos" meaning whole or entire and "graphein" meaning to write.

It is a two-step technique. First, an interference pattern between a wave field scattered from an object and a reference wave is recorded—the so-called hologram. The (two-dimensional) hologram contains the information about the entire three-dimensional wave field, coded in terms of interference stripes. The second step is the reconstruction of the object wave by illuminating with the reference wave again, which creates a three-dimensional image.

In *digital* holography, the hologram is recorded digitally on a charge-coupled device (CCD), and reconstructed numerically. Digital hologram recording was first performed by Schnars and Jüptner at Bremen University in 1993 [93]. There are numerous applications of digital holography, e.g. in material engineering and biological imaging. In this thesis we deal with another application: three-dimensional particle tracking in fluid mechanics. Here, the size and the distribution of particles are extracted from a hologram. In figure 16 the setup for particle digital holography is visualized.

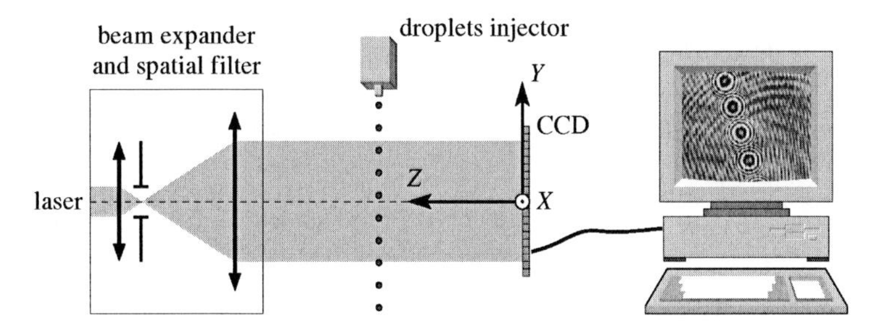

Figure 16: Setup for particle digital holography, from [97].

5.3.2 Data model

In digital holography, the data correspond to the diffraction patterns of the objects [39, 59]. Under Fresnel's approximation, diffraction can be modeled by a convolution with a "chirp" kernel. In the context of holograms of particles [47,48,105], the objects can be considered opaque (i.e. binary), and the hologram recorded on the camera corresponds to the convolution of disks with Fresnel's chirp kernels. The measurement of particle size and location therefore amounts to an inverse convolution problem [26, 97, 98].

We consider the case of spherical particles, which is of significant interest in applications such as fluid mechanics [77, 110]. We model the particles $j \in \{1, \ldots, N\}$ as opaque disks $B_r(\cdot - x_j, \cdot - y_j, \cdot - z_j)$ with center $(x_j, y_j, z_j) \in \mathbb{R}^3$, radius r and disk orientation orthogonal to the optical axis (Oz). Hence the source $f^\diamond$ is given as a sum of characteristic functions

$$f^\diamond = \sum_{j=1}^{N} u_j^\diamond \, \chi_{B_r}(\cdot - x_j, \cdot - y_j, \cdot - z_j) =: \sum_{j=1}^{N} u_j^\diamond \, \chi_j.$$

The real values $u_j^\diamond$ are amplitude factors of the diffraction pattern that in praxis depend on experimental parameters, cf. [97, 104].

To an incident laser beam of (complex) amplitude W_0 and wavelength λ, the amplitude W in the observation plane, i.e. at depth $z = 0$, is well modeled by a bidimensional convolution $\circledast$ with respect to (x, y). In the following ι represents the imaginary unit. Let δ_{x_j, y_j} denote Dirac's peak located at (x_j, y_j) and let h_{z_j} constitute the Fresnel function defined by

$$h_{z_j}(x, y) = \frac{1}{\iota \lambda z_j} \exp\left(\iota \frac{\pi}{\lambda z_j} \|R\|^2 \right), \qquad \text{with } R := (x, y).$$

With that, the amplitude $W : \mathbb{R}^2 \to \mathbb{C}$ in the observation plane can be expressed by

$$W = W_0 \Big[1 - \sum_{j=1}^{N} u_j^\diamond \left(\chi_j \circledast h_{z_j} \circledast \delta_{x_j, y_j} \right) \Big].$$

Remark that $h_{z_j} \circledast \delta_{x_j, y_j}$ denotes the shifted Fresnel function.

One difficulty occurring at digital holography inverse problems is, that in praxis only the absolute value of W can be measured by the detector, and the phase gets lost. The measured intensity consequently arises as

$$g = |W|^2 = |W_0|^2 \Big[1 - 2 \sum_{j=1}^{N} u_j^\diamond \left(\chi_j \circledast \mathrm{Re}(h_{z_j}) \circledast \delta_{x_j,y_j} \right) \\ + \sum_{i=1}^{N} \sum_{j=1}^{N} u_i^\diamond \left(\chi_i \circledast h_{z_i} \circledast \delta_{x_i,y_i} \right) u_j^\diamond \left(\chi_j \circledast h_{-z_j} \circledast \delta_{x_j,y_j} \right) \Big].$$

Since the second term is dominant over the third one for small χ, the intensity is classically linearized [97, 104]:

$$g \approx |W_0|^2 \Big[1 - 2 \sum_{j=1}^{N} u_j^\diamond \left(\chi_j \circledast \mathrm{Re}(h_{z_j}) \circledast \delta_{x_j,y_j} \right) \Big]. \tag{5.6}$$

The equation (5.6), i.e. the operator equation

$$A : f^\diamond = \sum u_j^\diamond \chi_j \mapsto g,$$

is the model for the inverse problem of convolution type, that appears in digital holography.

5.3.3 Resolution bounds for digital holography

Analogously to section 5.2, we next derive the Neumann εERCs for equation (5.6). For fixed (x_j, y_j, z_j) the associated (not necessarily unit-normed) atoms $\widetilde{\varphi}_{z_j} \in \Phi$ have the form

$$\widetilde{\varphi}_{z_j}(\cdot - x_j, \cdot - y_j) := \chi_{B_r}(\cdot - x_j, \cdot - y_j) \circledast \mathrm{Re}(h_{z_j}) \circledast \delta_{x_j,y_j}. \tag{5.7}$$

As for mass spectrometry in section 5.2, the first step is to calculate the norm of an atom and the correlation of two distinct ones. All calculations are done in $L^2(\mathbb{R}^2)$, i.e. the operator maps $A : L^2(\mathbb{R}^2) \to L^2(\mathbb{R}^2)$. Firstly, we need some properties of the Fresnel function.

Proposition 5.6. *For the convolution of the real part of the Fresnel function we have the following properties:*

$$\begin{aligned} Re(h_{z_1}) \circledast Re(h_{z_2}) &= \tfrac{1}{2}\left(Re(h_{z_1+z_2}) + Re(h_{z_1-z_2}) \right), && \text{for all } z_1, z_2 \in \mathbb{R}, \\ Re(h_z) \circledast Re(h_z) &= \tfrac{1}{2}\left(\delta + Re(h_{2z}) \right), && \text{for all } z \in \mathbb{R}. \end{aligned}$$

Proof. Due to [63], for the Fresnel function it holds that

$$h_{z_1} \circledast h_{z_2} = h_{z_1+z_2}, \qquad \text{for all } z_1, z_2 \in \mathbb{R},$$
$$h_z \circledast h_{-z} = \delta, \qquad \text{for all } z \in \mathbb{R}.$$

With that and since $\overline{h_z} = h_{-z}$ the statement follows. □

Another property that is required for the deduction of the Neumann εERC is, that the convolution of a function with the Fresnel function—the so-called Fresnel transform—can be related to a direct multiplication with its Fourier transform which is defined by

$$\mathcal{F}\phi(\xi,\nu) := \int\limits_{\mathbb{R}^2} \phi(x,y) \exp(-2\pi\iota(x\xi + y\nu)) \,\mathrm{d}x \,\mathrm{d}y.$$

Proposition 5.7. *Let $\phi \in L^2(\mathbb{R}^2)$ and h_z be a Fresnel function. Then*

$$(\phi \circledast h_z)(\xi,\nu) = \mathcal{F}\big\{\iota\lambda z\, h_z\, \phi\big\}\Big(\frac{\xi}{\lambda z}, \frac{\nu}{\lambda z}\Big)\, h_z(\xi,\nu).$$

Proof. Let $\phi \in L^2(\mathbb{R}^2)$, $z \in \mathbb{R}$ and h_z be the corresponding Fresnel function. Then rearranging yields to the statement,

$$\begin{aligned}
(\phi \circledast h_z)(\xi,\nu) &= \int\limits_{\mathbb{R}^2} \phi(x,y)\, \tfrac{1}{\iota\lambda z} \exp\big(\tfrac{\iota\pi}{\lambda z}((x-\xi)^2 + (y-\nu)^2)\big) \,\mathrm{d}x \,\mathrm{d}y \\
&= \tfrac{1}{\iota\lambda z} \exp\big(\tfrac{\iota\pi}{\lambda z}(\xi^2+\nu^2)\big) \\
&\qquad \int\limits_{\mathbb{R}^2} \phi(x,y) \exp\big(\tfrac{\iota\pi}{\lambda z}(x^2+y^2)\big) \exp\big(-2\pi\iota(\tfrac{x\xi}{\lambda z} + \tfrac{y\nu}{\lambda z})\big) \,\mathrm{d}x \,\mathrm{d}y \\
&= \mathcal{F}\big\{\iota\lambda z\, h_z\, \phi\big\}\big(\tfrac{\xi}{\lambda z}, \tfrac{\nu}{\lambda z}\big)\, h_z(\xi,\nu).
\end{aligned}$$

□

Remark 5.8. In praxis, the function ϕ has a bounded and small support with respect to $\sqrt{\lambda z}$. With $(x^2+y^2)_{\max}$ denoting the maximal spatial dimension of ϕ respectively the maximal spatial extend of the corresponding particle, the so-called far-field condition $\frac{(x^2+y^2)_{\max}}{\lambda z} \ll 1$ holds in the proof of proposition 5.7, cf. [104]. In [97] e.g., particles of radius at about 50µm are illuminated with a red laser beam (wavelength 630nm) and a distance to camera of about 250mm. Thus the term

$(x^2+y^2)_{\max}/(\lambda z) \approx 3\cdot 10^{-4}$ and hence $\exp\left(\frac{\iota\pi(x^2+y^2)}{\lambda z}\right)$ is approximatively 1. On the far-field condition, we can estimate

$$\big(\phi \circledast h_z\big)(\xi,\nu) \approx \mathcal{F}\phi\Big(\frac{\xi}{\lambda z},\frac{\nu}{\lambda z}\Big)\, h_z(\xi,\nu). \tag{5.8}$$

With that, for the complex valued diffraction, with $\varrho := (\xi,\nu)$ and J_1 denoting the first kind Bessel function of order one, we get

$$\big(\chi_{B_r} \circledast h_z\big)(\varrho) \approx \frac{r}{\iota\|\varrho\|}\, J_1\Big(\frac{2\pi r}{\lambda z}\|\varrho\|\Big)\ \exp\Big(\iota\frac{\pi}{\lambda z}\|\varrho\|^2\Big),$$

since $\mathcal{F}\chi_{B_r}(\varrho) = 2\pi r^2\left[\frac{J_1(2\pi r\|\varrho\|)}{2\pi r\|\varrho\|}\right]$ holds (Airy's pattern, vide infra). With that, for a real valued intensity atom we get

$$\begin{aligned}\widetilde{\varphi}_z(\varrho) &= \big(\chi_{B_r} \circledast \mathrm{Re}(h_z)\big)(\varrho) = \big(\mathrm{Re}(\chi_{B_r} \circledast h_z)\big)(\varrho)\\ &\approx \tfrac{r}{\|\varrho\|} J_1\Big(\tfrac{2\pi r}{\lambda z}\|\varrho\|\Big) \sin\Big(\tfrac{\pi}{\lambda z}\|\varrho\|^2\Big),\end{aligned} \tag{5.9}$$

which corresponds to the model given by Tyler and Thompson in [104].

Back to the correlation and—as a special case—the norm of an atom. The correlation appears as the autoconvolution, namely,

$$\begin{aligned}&\big\langle\widetilde{\varphi}_{z_i}(\cdot - x_i, \cdot - y_i), \widetilde{\varphi}_{z_j}(\cdot - x_j, \cdot - y_j)\big\rangle_{L^2}\\ &\qquad= \int_{\mathbb{R}^2} \widetilde{\varphi}_{z_i}(x,y)\, \widetilde{\varphi}_{z_j}(x-(x_j-x_i), y-(y_j-y_i))\ \mathrm{d}x\ \mathrm{d}y\\ &\qquad= \big(\widetilde{\varphi}_{z_i} \circledast \widetilde{\varphi}_{z_j}\big)(x_j-x_i, y_j-y_i).\end{aligned}$$

In the following we assume that all particles are located in a plane parallel to the detector, i.e. $z := z_i$ is constant for all i. Then the autoconvolution of an atom appears as

$$\widetilde{\varphi}_z \circledast \widetilde{\varphi}_z = \chi_{B_r} \circledast \chi_{B_r} \circledast \mathrm{Re}(h_z) \circledast \mathrm{Re}(h_z).$$

Let $\varrho := (\xi,\nu)$. With proposition 5.6 and the formula

$$\begin{aligned}C(\varrho) :&= \big(\chi_{B_r} \circledast \chi_{B_r}\big)(\varrho)\\ &= \begin{cases} 2r^2\cos^{-1}\left(\frac{\|\varrho\|}{2r}\right) - \frac{\|\varrho\|}{2}\sqrt{4r^2-\|\varrho\|^2}, & \text{for } 4r^2 > \|\varrho\|^2,\\ 0, & \text{else,}\end{cases}\end{aligned}$$

we get

$$\widetilde{\varphi}_z \circledast \widetilde{\varphi}_z = C \circledast \tfrac{1}{2}\Big[\delta + \operatorname{Re}(h_{2z})\Big] = \tfrac{1}{2}\Big[C + C \circledast \operatorname{Re}(h_{2z})\Big]. \tag{5.10}$$

With equation (5.8) and since $\mathcal{F}C$ is real valued, we get

$$\begin{aligned} C \circledast \operatorname{Re}(h_{2z}) &= \operatorname{Re}(C \circledast h_{2z}) \approx \operatorname{Re}\Big(\mathcal{F}C(\cdot/\lambda z)\, h_{2z}\Big) = \mathcal{F}C(\cdot/\lambda z) \operatorname{Re}(h_{2z}) \\ &= \mathcal{F}\chi_{B_r}(\cdot/\lambda z)\, \mathcal{F}\chi_{B_r}(\cdot/\lambda z) \operatorname{Re}(h_{2z}). \end{aligned}$$

In physics, it is well known that the Fourier transform of a disc is the Bessel cardinal function, defined by $\operatorname{Jinc}(x) := J_1(x)/x$, since it is the diffraction of a circular aperture at infinite distance, cf. [1, remark 2.1]. Nevertheless, for the sake of completeness and by reasons of mathematical beauty we illustrate this computation. Since the Fourier transform of a radial function is the Hankel transform of order zero (also known as Bessel transform of order zero), cf. [99, Theorem IV.3.3, page 155], the Fourier transform of χ_{B_r} appears for $\varrho := (\xi, \nu)$ as

$$\mathcal{F}\chi_{B_r}(\varrho) = 2\pi \int_0^r S\, J_0(2\pi\|\varrho\|S)\ \mathrm{d}\,S = \frac{1}{2\pi\|\varrho\|^2} \int_0^{2\pi r\|\varrho\|} S\, J_0(S)\ \mathrm{d}\,S.$$

In order to solve this definite integral, we use $\int SJ_0(S)\ \mathrm{d}\,S = SJ_1(S)$, cf. [51, equation 5.52 1.], and get

$$\mathcal{F}\chi_{B_r}(\varrho) = 2\pi r^2 \Big[\frac{J_1(2\pi r\|\varrho\|)}{2\pi r\|\varrho\|}\Big],$$

hence the Fourier transform of the circle-circle intersection C appears as

$$\mathcal{F}C(\varrho) = \mathcal{F}\big(\chi_{B_r} \circledast \chi_{B_r}\big)(\varrho) = \mathcal{F}\chi_{B_r}(\varrho)\, \mathcal{F}\chi_{B_r}(\varrho) = \tfrac{r^2}{\|\varrho\|^2} J_1^2(2\pi r\|\varrho\|).$$

With that result, we can easily calculate the norm of an atom $\widetilde{\varphi}_z$. Since $C(0) = \pi r^2$, $\mathcal{F}C(0) = \big(\mathcal{F}\chi_{B_r}(0)\big)^2 = \big(\int \chi_{B_r}\ \mathrm{d}\,x\big)^2 = \pi^2 r^4$ and $h_{2z}(0) = 0$ we obtain

$$\|\widetilde{\varphi}_z\|_{L^2}^2 = \big|\widetilde{\varphi}_z \circledast \widetilde{\varphi}_z\big|(0) \approx \tfrac{1}{2}\pi r^2.$$

Hence, for fixed z we can represent the associated unit-normed atoms $\varphi_z \in \Phi$ with $R := (x, y)$ via

$$\varphi_z := \frac{\widetilde{\varphi}_z}{\|\widetilde{\varphi}_z\|} \approx \Big(\frac{2}{\pi}\Big)^{\frac{1}{2}} \frac{1}{\|R\|} J_1\Big(\frac{2\pi r}{\lambda z}\|R\|\Big) \sin\Big(\frac{\pi}{\lambda z}\|R\|^2\Big). \tag{5.11}$$

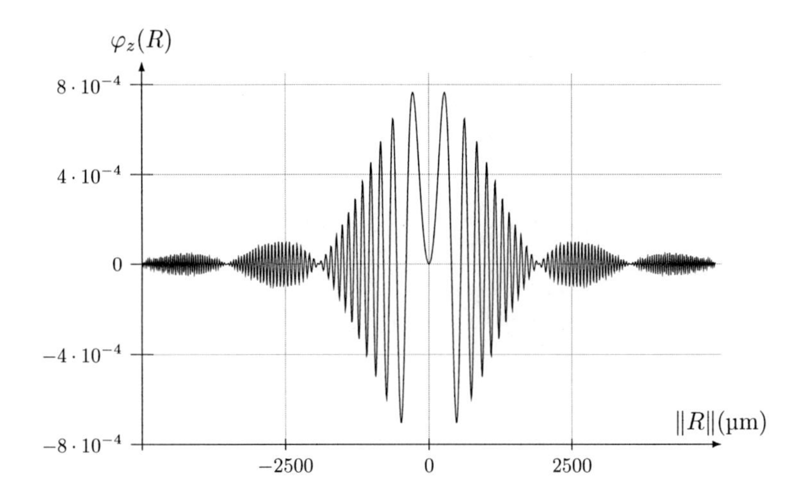

Figure 17: Unit-normed, centered atom of particles of radius 50µm, illuminated with a red laser beam (wavelength 630nm) and distance to camera of 250mm.

In figure 17 the centered atom for a particle of 50µm radius is displayed, which is illuminated with a red laser beam (wavelength 630nm) in a distance of 250mm to the camera.

The autoconvolution for general ϱ and hence the correlation of two atoms $\varphi_z(\cdot - x_i, \cdot - y_i)$ and $\varphi_z(\cdot - x_j, \cdot - y_j)$ with distance distance $\varrho_{j,i} := (x_j - x_i, y_j - y_i)$ in digital holography emerges as

$$\begin{aligned}
\Big|\big\langle \varphi_z(\cdot - x_i, \cdot - y_i), \varphi_z(\cdot - x_j, \cdot - y_j)\big\rangle\Big| &= \big|\varphi_z \circledast \varphi_z\big|(\varrho_{j,i}) \\
&= \frac{1}{\|\widetilde{\varphi}_z\|_{L^2}^2} \big|\widetilde{\varphi}_z \circledast \widetilde{\varphi}_z\big|(\varrho_{j,i}) \\
&\approx \frac{1}{\pi r^2} \Big[C(\varrho_{j,i}) + \mathcal{F}C\Big(\frac{\varrho_{j,i}}{\lambda z}\Big) \big|\mathrm{Re}(h_{2z}(\varrho_{j,i}))\big|\Big] \\
&= \frac{C(\varrho_{j,i})}{\pi r^2} + \frac{1}{4} J_1^2\Big(\frac{2\pi r}{\lambda z}\|\varrho_{j,i}\|\Big) \Big|\operatorname{sinc}\Big(\frac{1}{2\lambda z}\|\varrho_{j,i}\|^2\Big)\Big|, \qquad (5.12)
\end{aligned}$$

where sinc denotes the normalized sine cardinal function and is defined via $\operatorname{sinc}(x) := \sin(\pi x)/\pi x$.

The correlation in digital holography (5.12) is not as easily valuable as in mass spectrometry, because it is not monotonically decreasing in the distance $\|\varrho_{j,i}\| = \sqrt{(x_j - x_i)^2 + (y_j - y_i)^2}$ due to the oscillating Bessel and sine functions. To come to an estimate from above, which is monotonically decreasing, we use bounds for the absolute value of the Bessel functions J_1^2. In [60], the author gives estimates for $|J_\nu(x)|$ for $x > 0$ and $\nu > 0$, namely,

$$|J_\nu(x)| \leq \min\{b_L \nu^{-1/3}, c_L x^{-1/3}\}, \tag{5.13}$$

with constants

$$b_L := \sqrt[3]{2} \sup_{x>0} \tfrac{\sqrt{x}}{3}\left(J_{-\frac{1}{3}}\left(\tfrac{2}{3}x^{\frac{3}{2}}\right) + J_{\frac{1}{3}}\left(\tfrac{2}{3}x^{\frac{3}{2}}\right)\right) \approx 0.6748,$$
$$c_L := \sup_{x>0} x^{\frac{1}{3}} J_0(x) \approx 0.7857.$$

In addition, the sine cardinal function obviously is bounded from above via 1 and $1/x$, and hence we have

$$|\varphi_z \circledast \varphi_z|(\varrho) \leq \frac{C(\varrho)}{\pi r^2} + \frac{1}{4} \min\left\{b_L^2, c_L^2\left(\frac{\lambda z}{2\pi r}\right)^{\frac{2}{3}} \|\varrho\|^{-\frac{2}{3}}\right\} \min\left\{1, \frac{2\lambda z}{\pi}\|\varrho\|^{-2}\right\}, \tag{5.14}$$

which is monotonically decreasing in $\|\varrho\|$. Figure 18 illustrates the oscillating part of the correlation (5.12) and its corresponding upper bound from (5.14) for two particles of 50µm radius, which are illuminated with a red laser beam (wavelength 630nm) in a distance of 250mm to the camera. For the sake of simplicity we denote the above majorizing function by $\mathcal{M}$, i.e.

$$\mathcal{M}(\varrho) := \frac{C(\varrho)}{\pi r^2} + \frac{1}{4} \min\left\{b_L^2, c_L^2\left(\frac{\lambda z}{2\pi r}\right)^{\frac{2}{3}} \|\varrho\|^{-\frac{2}{3}}\right\} \min\left\{1, \frac{2\lambda z}{\pi}\|\varrho\|^{-2}\right\}. \tag{5.15}$$

Remark 5.9. In [58], the author gives a sharper estimation for $J_\nu^2(x)$, namely, for $\nu > -1/2$, $\varsigma := (2\nu+1)(2\nu+3)$ and $x > \sqrt{\varsigma + \varsigma^{2/3}}/2$,

$$J_\nu^2(x) \leq \frac{4(4x^2 - (2\nu+1)(2\nu+5))}{\pi((4x^2 - \varsigma)^{3/2} - \varsigma)}.$$

With that (asymptotically $|\varphi_z \circledast \varphi_z|(\varrho) \sim \|\varrho\|^{-3}$) instead of the rough bound (5.13) (asymptotically $|\varphi_z \circledast \varphi_z|(\varrho) \sim \|\varrho\|^{-\frac{8}{3}}$), it is possible to get a more precise recovery condition for digital holography. Since this technical computation is beyond the scope of this theoretical section, we postpone it here.

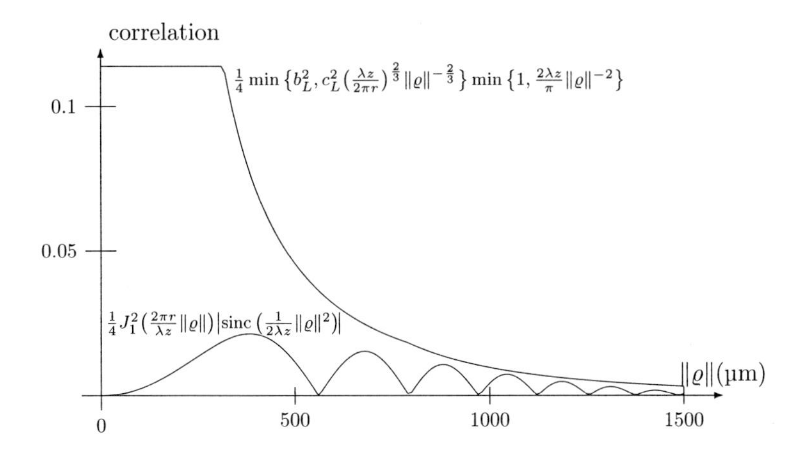

Figure 18: The oscillating part of the correlation of two atoms with distance $\|\varrho\|$ and its corresponding monotonically decreasing estimate (same settings as in figure 17).

With the estimation (5.14), we come to a resolution bound for droplets jet reconstruction, as e.g. used in [97]. Here monodisperse droplets (i.e. they have the same size, shape and mass) were generated and emitted on a strait line parallel to the detector plane, see figure 16, page 97. This configuration eases the computation of the εNeumann ERCs. Analogously to the mass spectrometry example in section 5.2, we define that the particles are located at some points of the grid

$$\Delta\mathbb{Z} := \{i \in \mathbb{Z} \,|\, i/\Delta \in \mathbb{Z}\},$$

where the parameter Δ describes the dictionary refinement. Assume that the particles have the minimal distance

$$\rho := \min_{i,j\in \mathrm{supp}(u^\diamond)} \|\varrho_{j,i}\| \in \Delta\mathbb{N},$$

then the sums of correlations COR_I and $\mathrm{COR}_{I^\complement}$ can be estimated from above. For $\rho > \Delta$ we get

$$\mathrm{COR}_I = \sup_{i\in I} \sum_{\substack{j\in I\\ j\neq i}} |\langle \varphi_i, \varphi_j\rangle| \leq 2 \sum_{j=1}^{\lfloor N/2\rfloor} \mathcal{M}(j\rho),$$

$$\mathrm{COR}_{I^\complement} = \sup_{i\in I^\complement} \sum_{j\in I} |\langle \varphi_i, \varphi_j\rangle| \leq \sup_{\substack{i\in\Delta\mathbb{Z}\\ \Delta\leq i\leq \rho-\Delta}} \sum_{j=-\lfloor N/2\rfloor}^{\lfloor N/2\rfloor} \mathcal{M}(|j\rho - i|).$$

Remark 5.10. Same as before for mass spectrometry, if the cardinality N of the support I is unknown then one can replace the finite sums by infinite sums. These sums exist and can be expressed in terms of the Hurwitz zeta function $\zeta(\nu, q) := \sum_{j=0}^{\infty}(q+j)^{-\nu}$, for $\nu > 1$ and $q > 0$, and the Riemann zeta function $\zeta(\nu) := \zeta(\nu, 1) = \sum_{j=1}^{\infty} j^{-\nu}$, respectively.

With that, we can formulate an estimation for the Neumann εERCs for the ℓ^1-penalized Tikhonov regularization (3.26) and OMP (4.9).

Proposition 5.11 (Neumann εERC for Fresnel-convolved characteristic functions)**.** *An estimation from above for the Neumann εERCs (3.26) and (4.9) for characteristic functions convolved with the real part of the Fresnel kernel is for $\rho > \Delta$*

$$2\sum_{j=1}^{\lfloor N/2\rfloor} \mathcal{M}(j\rho) + \sup_{1\leq i<\frac{\rho}{\Delta}} \sum_{j=-\lfloor N/2\rfloor}^{\lfloor N/2\rfloor} \mathcal{M}(|j\rho - i\Delta|) < 1 - 2r_{\varepsilon/u}. \quad (5.16)$$

This means, that there is a regularization parameter α which allows exact recovery of the support with the ℓ^1-penalized Tikhonov regularization, and that we are able to recover the support with OMP exactly, if the above condition is fulfilled.

Condition (5.16) of proposition 5.11 seems not to be easily to handle due to the upper bound $\mathcal{M}$ from (5.15). However, in praxis all parameters are known, and one can compute a bound via approaching from large ρ. As soon as the sum is smaller than $1-2r_{\varepsilon/u}$, it is guaranteed that both, the ℓ^1-penalized Tikhonov regularization and OMP, can recover exactly. A typical setting for digital holography of particles is the usage of a red laser of wavelength $\lambda = 0.6328\mu\mathrm{m}$ and a distance of $z = 200\mathrm{mm}$ from the camera, cf. [97]. In figure 19 condition (5.16) of proposition 5.11

is plotted for particles with typical radii of $r \in \{5, 25, 50, 100\}$µm. In the computation the asymptotic formula is used, i.e. for an unknown support cardinality N. For the dictionaries a corresponding refinement of $\Delta = r/2$ was chosen. The colored areas describe the combinations where the Neumann εERC (5.16) is fulfilled, and hence exact recovery is possible.

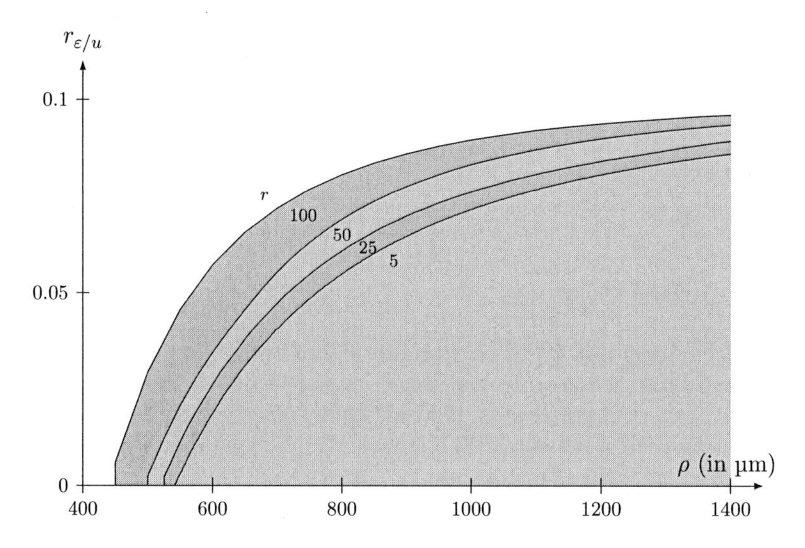

Figure 19: The Neumann εERC (5.16) for combinations of ρ and $r_{\varepsilon/u}$ with the asymptotic formula for an unknown support cardinality N. For particles the radii $r \in \{5, 25, 50, 100\}$µm were used, and for the corresponding dictionary we choosed a refinement of $\Delta = r/2$.

5.3.4 Numerical examples

We apply the Neumann εERC (5.16) to simulated data of droplets jets. For the simulation we use the same setting as above, i.e. a red laser of wavelength $\lambda = 0.6328$µm and a distance of $z = 200$mm from the camera. The particles have a diameter of 100µm and for the corresponding dictionary we choose a refinement of 25µm. Those parameters correspond to that of the experimental setup used in [97, 98].

After applying the digital holography model (5.6), we add Gaussian noise of different noise levels and in each case of zero mean. For the coefficients $u_i^\diamond$, we choose a setting which implies $u_i^\diamond \approx 10$ for all $i \in I$. Figure 20 shows three simulated holograms with different distances ρ and different noise-to-signal ratios $r_{\varepsilon/u}$. For all three noisy examples in the right column all the particles were recovered exactly, both with ℓ^1-penalized Tikhonov regularization and with the orthogonal matching pursuit. For minimization of the Tikhonov functional we used the iterated soft-thresholding algorithm [25]. However, only for the image on top ($\rho \approx 721$µm) condition (5.16) of proposition 5.11 holds. In the image in the middle of figure 20, the particles have a too small distance to each other ($\rho \approx 360$µm), and even for the noiseless case condition (5.16) is not fulfilled. The last image ($\rho \approx 721$µm) was manipulated with unrealistically huge noise, so that condition (5.16) is violated, too.

5.4 Conclusion

In this chapter we have showed the practical relevance of the Neumann εERCs for the ℓ^1-penalized Tikhonov regularization (3.26) and OMP (4.9) with two real-world applications. The motivation of the deduction of the Neumann εERCs in chapters 3 and 4 has been to treat ill-posed problems, and in particular, these two problems of convolution type.

To obtain a priori computable conditions, a main tool in this chapter has been, that the atoms in the dictionary are shifted copies of the same shape, and that the correlation of the atoms depends on the distance of the atoms only. Once there is a sufficiently decaying upper bound for the correlation, we are able to apply the Neumann εERC and obtain a priori computable conditions for exact recovery. In the two examples considered here, we have evaluated the correlation of two atoms analytically. If this is not possible, one can alternatively derive the correlation numerically and determine a numerical envelope of the correlation, which is monotonically decreasing in the distance of the atoms.

Especially in the example from digital holography, the analysis may be regarded as a first step to calculate the resolution power of droplet holography. However, the experiments indicate that the conditions for exact recovery from propositions 3.17 and 4.10 are too restrictive.

An idea to come to tighter exact recovery conditions is to bring in more prior knowledge, as e.g. a nonnegativity constraint. We postpone this idea for future work. For digital holography even more prior knowledge

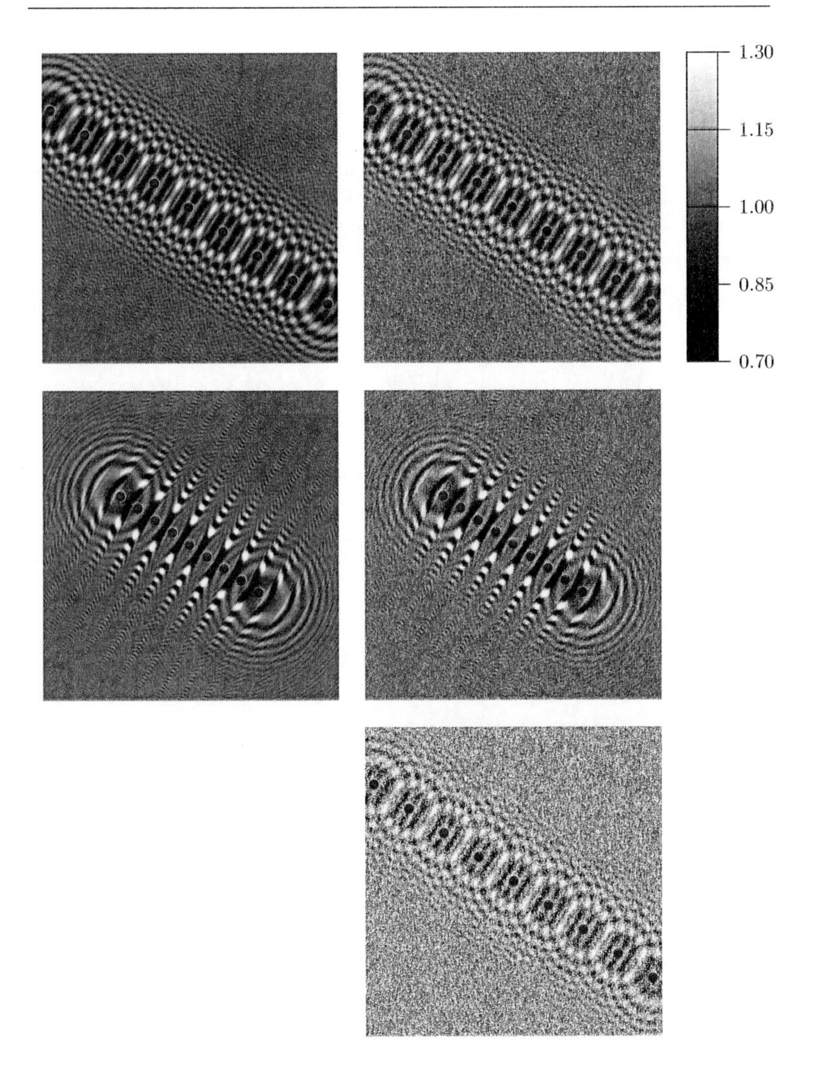

Figure 20: Simulated holograms of spherical particles. In the left column the noiseless signals are displayed. For reconstruction, the noisy signals of the right column are used. The dots correspond to the true location of the particles. Both, the ℓ^1-penalized Tikhonov regularization and the orthogonal matching pursuit recovered all particles exactly, however, condition (5.16) of proposition 5.11 was just fulfilled for the image on top. In the image in the middle the particles have a too small distance to each other, and at the bottom the image was manipulated with unrealistically huge noise.

may be taken into account, since the objects are not only nonnegative, but even all apertures have the same denseness, i.e. the coefficients u_i are constant for all $i \in I$.

As discussed in remark 5.5, a straightforward generalization of our approach to fully continuous dictionaries runs into problems. Maybe one may obtain bounds on how accurate the support is localized. This is strongly related to the structure of the dictionary (e.g. that is consists of shifts of the same object) and of course related to the correlations. In particular, for Gaussian-convolved Dirac peaks the analysis of errors in the Prokhorov metric is a further direction of research.

For the example from digital holography of particles we so far have restricted the setting to particles located in a plane parallel to the detector, and we firstly have given resolution bounds for droplets jets, i.e. the particles are positioned at a one-dimensional grid. Furthermore, it is of interest to give conditions in a two- or three-dimensional setting. We postpone this idea for future work.

C Conclusion

This thesis contributes to regularization theory of inverse problems with sparsity constraints. We have considered the linear operator equation

$$Af = g,$$

where only noisy data g^ε are available. For the unknown solution f we have assumed, that it can be represented using only a few elements of a suitable basis or dictionary $\Psi = \{\psi_i\}_{i\in\mathbb{Z}}$, i.e.

$$f = \sum_i u_i \psi_i,$$

where the cardinality of the support $I := \operatorname{supp}(u)$ is small. In the previous chapters, for the Tikhonov regularization with a sparsity enforcing penalty and for the orthogonal matching pursuit we have obtained a couple of stability results, namely, convergence rates and exact recovery conditions.

In chapter 2, the Tikhonov regularization with a sparsity-enforcing penalty has been analyzed in scales of Banach spaces, namely, in the scale of Besov spaces. The results generalize regularization in Hilbert scales initiated in [78]. We have used the Besov scale to model smoothing properties of the operator A, the regularization term, and the source condition. The main result of this chapter is the fact that, on the one hand, tighter source conditions may not lead to stronger convergence rates, and, on the other hand, a less tight source condition may lead to a stronger result. This is different to classical regularization in Hilbert scales. The examples in chapter 2 have shown only slight improvements, and it is questionable if the effect can be observed numerically. However, the effect that looser source conditions lead to tighter convergence results is interesting on its own.

In chapter 3, the Tikhonov regularization with an ℓ^1 penalty has been analyzed. We have seen that this regularization method enforces a sparse approximate solution. In [41], the authors introduce an a priori parameter rule that ensures linear convergence to the minimum-norm solution. In this thesis we have presented a parameter rule which ensures exact recovery of the unknown support I of u. We have introduced the *Neumann exact recovery condition* which is sufficient for the existence of a regularization parameter that provides exact recovery. The condition depends on inner products of images of the operator A, namely, on

$$\mathrm{COR}_I = \sup_{i\in I} \sum_{\substack{j\in I\\ j\neq i}} |\langle A\psi_i, A\psi_j\rangle| \quad \text{and} \quad \mathrm{COR}_{I^\complement} = \sup_{i\in I^\complement} \sum_{j\in I} |\langle A\psi_i, A\psi_j\rangle|.$$

In order to use this parameter rule, one has to know the exact support I. In this case there would be no need to apply complex reconstruction methods. However, with a certain prior knowledge one can estimate COR_I and $\mathrm{COR}_{I^\complement}$ from above and obtain an a priori computable parameter rule for exact recovery.

In chapter 4, convergence rates and conditions for exact recovery for the orthogonal matching pursuit (OMP) have been proven. We have deduced a recovery condition that also depends on COR_I and $\mathrm{COR}_{I^\complement}$, and which is called Neumann exact recovery condition, too. If the operator A is unit normed, i.e. if $\|A\psi_i\| = 1$ for all $i \in \mathbb{Z}$, then the condition coincides with the Neumann condition for the Tikhonov regularization deduced in chapter 3. Moreover, there is a simple error bound for approximate solutions determined by OMP, which shows a convergence rate of $\mathcal{O}(\varepsilon)$. The rate of convergence resembles what is known for the Tikhonov regularization with an ℓ^1-penalty term.

Finally, in chapter 5 the practical relevance of the Neumann exact recovery conditions has been demonstrated. The conditions have been used to obtain a priori computable conditions for two examples of convolution type, first, for impulse trains convolved with a Gaussian kernel as it e.g. occurs in mass spectrometry, and second, for characteristic functions convolved with a Fresnel kernel as it is e.g. used in digital holography of particles. In each example we have normalized the operator A, so that the conditions for the ℓ^1-penalized Tikhonov functional and for OMP are equal.

In these two real-world applications we have shown that the Neumann conditions lead to a priori computable conditions. A main tool here has been, that the atoms of the dictionary $\{A\psi_i\}_{i\in\mathbb{Z}}$ are shifted copies of the same shape, and that the correlation of the atoms depends on the distance of the atoms only. Once there is a sufficiently decaying upper bound for the correlation, we are able to apply the Neumann condition and obtain a priori computable conditions for exact recovery.

There are many points for further investigation. We have expatiated on them in the concluding sections of chapters 2 to 5. Nevertheless, some important aspects should be repeated here.

In chapter 2, we have restricted the analysis to a Banach-to-Hilbert space setting, namely, to operators mapping from a Besov space $\mathcal{B}_D$ to the Hilbert space L^2. Assuming an operator $A : \mathcal{B}_D \to \mathcal{B}_E$ with a certain Besov space $\mathcal{B}_E$ may lead to more general results in a Banach-to-Banach space setting. This analysis is of interest.

For both methods, the Tikhonov regularization with an ℓ^1 penalty and the orthogonal matching pursuit, the Neumann exact recovery condition gives a deterministic statement for exact recovery. However, in a lot of applications it is not possible to obtain exact recovery for all signals but for almost all, i.e. the exact recovery happens with a high propability. Hence, hoping for definite exact recovery is too optimistic. The deduction of probabilistic statements possibly could lead to looser conditions that are more applicable to real-life data.

The analysis of the example from digital holography may be regarded as a first step to calculate the resolution power of droplet holography. However, there are many questions remained open. The following ones should be mentioned. We so far have restricted ourselves to a one-dimensional setting. The deduction of resolution bounds in a two- or three-dimensional setting is of high practical relevance, in particular to give the axial resolution that is achievable. Furthermore, the experiments in section 5.3 indicate that the Neumann exact recovery condition for digital holography of particles is too restrictive. Since the objects can be considered opaque (i.e. binary), the coefficients u_i are constant for all $i \in I$. It may be possible to achieve tighter exact recovery conditions by taking into account this prior knowledge. This analysis is of high practical importance, as well.

Bibliography

[1] M. Bertero and P. Boccacci. *Introduction to Inverse Problems in Imaging*. Institute of Physics Publishing, 1998.

[2] J. D. Blanchard, C. Cartis, J. Tanner, and A. Thompson. Phase transitions for greedy sparse approximation algorithms. Technical Report ERGO 09-010, School of Mathematics, University of Edinburgh, 2009.

[3] T. Bonesky. Morozov's discrepancy principle and Tikhonov-type functionals. *Inverse Problems*, 25(1):015015 (11pp), 2009.

[4] T. Bonesky, K. S. Kazimierski, P. Maass, F. Schöpfer, and T. Schuster. Minimization of Tikhonov functionals in Banach spaces. *Abstract and Applied Analysis*, vol. 2008:Art. ID 192679, 19 pages, 2008.

[5] K. Bredies. A forward-backward splitting algorithm for the minimization of non-smooth convex functionals in Banach space. *Inverse Problems*, 25(1):015005 (20pp), 2009.

[6] K. Bredies, T. Alexandrov, J. Decker, D. A. Lorenz, and H. Thiele. Sparse deconvolution for peak picking and ion charge estimation in mass spectrometry. In *Proceedings of the 15th European Conference on Mathematics for Industry*, 2008.

[7] K. Bredies and D. A. Lorenz. Iterated hard shrinkage for minimization problems with sparsity constraints. *SIAM Journal on Scientific Computing*, 30(2):657–683, 2008.

[8] K. Bredies and D. A. Lorenz. Linear convergence of iterative soft-thresholding. *Journal of Fourier Analysis and Applications*, 14(5–6):813–837, 2008.

[9] K. Bredies and D. A. Lorenz. Regularization with non-convex separable constraints. *Inverse Problems*, 25(8):085011 (14pp), 2009.

[10] A. M. Bruckstein, D. L. Donoho, and M. Elad. From sparse solutions of systems of equations to sparse modeling of signals and images. *SIAM Review*, 51(1):34–81, 2009.

[11] A. M. Bruckstein, M. Elad, and M. Zibulevsky. A non-negative and sparse enough solution of an underdetermined linear system of equations is unique. *IEEE Transactions on Information Theory*, 54(11):4813–4820, 2008.

[12] M. Burger and S. Osher. Convergence rates of convex variational regularization. *Inverse Problems*, 20(5):1411–1420, 2004.

[13] E. J. Candés and J. K. Romberg. Sparsity and incoherence in compressive sampling. *Inverse Problems*, 23(3):969–985, 2007.

[14] E. J. Candés, J. K. Romberg, and T. Tao. Stable signal recovery from incomplete and inaccurate measurements. *Communications on Pure and Applied Mathematics*, 59(8):1207–1223, 2006.

[15] E. J. Candés and T. Tao. Decoding by linear programming. *IEEE Transaction on Information Theory*, 51(12):4203–4215, 2005.

[16] E. J. Candés and M. B. Wakin. An introduction to compressive sampling. *IEEE Signal Processing Magazine*, 25(2):21–30, 2008.

[17] R. Caprioli, T. Farmer, and J. Gile. Molecular imaging of biological samples : Localization of peptides and proteins using MALDI-TOF MS. *Analytical Chemistry*, 69(23):4751–4760, 1997.

[18] A. Chambolle, R. A. DeVore, N. Lee, and B. J. Lucier. Nonlinear wavelet image processing: Variational problems, compression and noise removal through wavelet shrinkage. *IEEE Transactions on Image Processing*, 7(3):319–335, 1998.

[19] T. F. Chan and J. Shen. *Image Processing and Analysis - Variational, PDE, Wavelet, and Stochastic Methods.* SIAM, Philadelphia, 2005.

[20] S. S. Chen, D. L. Donoho, and M. A. Saunders. Atomic decomposition by basis pursuit. *SIAM Journal on Scientific Computing*, 20(1):33–61, 1998.

[21] A. Cohen. *Numerical Analysis of Wavelet Methods.* Elsevier Science B. V., 2003.

[22] S. Dahlke, M. Fornasier, and T. Raasch. Multilevel preconditioning for adaptive sparse optimization. Preprint 25, DFG SPP 1324, 2009.

[23] S. Dahlke, P. Maass, G. Teschke, K. Koch, D. A. Lorenz, S. Müller, S. Schiffler, A. Stämpfli, H. Thiele, and M. Werner. *Mathematical Methods in Time Series Analysis and Digital Image Processing*, chapter Multiscale Approximation, pages 75–110. Understanding Complex Sytems. Springer, 2007.

[24] W. Dai and O. Milenkovic. Subspace pursuit for compressive sensing signal reconstruction. *IEEE Transactions on Information Theory*, 55(5):2230–2249, 2009.

[25] I. Daubechies, M. Defrise, and C. De Mol. An iterative thresholding algorithm for linear inverse problems with a sparsity constraint. *Communications in Pure and Applied Mathematics*, 57(11):1413–1457, 2004.

[26] L. Denis, D. A. Lorenz, E. Thiébaut, C. Fournier, and D. Trede. Inline hologram reconstruction with sparsity constraints. *Optics Letters*, 34(22):3475–3477, 2009.

[27] L. Denis, D. A. Lorenz, and D. Trede. Greedy solution of ill-posed problems: Error bounds and exact inversion. *Inverse Problems*, 25(11):115017 (24pp), 2009.

[28] R. A. DeVore. Nonlinear approximation. *Acta Numerica*, pages 51–150, 1998.

[29] D. L. Donoho, M. Elad, and V. Temlyakov. Stable recovery of sparse overcomplete representations in the presence of noise. *IEEE Transactions on Information Theory*, 52(1):6–18, 2006.

[30] D. L. Donoho, Y. Tsaig, I. Drori, and J.-L. Starck. Sparse solution of underdetermined linear equations by stagewise orthogonal

matching pursuit. Technical Report 2006-02, Stanford, Department of Statistics, 2006.

[31] C. Dossal and S. Mallat. Sparse spike deconvolution with minimum scale. In *Proceedings of the First Workshop "Signal Processing with Adaptive Sparse Structured Representations"*, 2005.

[32] H. W. Engl, M. Hanke, and A. Neubauer. *Regularization of Inverse Problems*, volume 375 of *Mathematics and its Applications*. Kluwer Academic Publishers Group, Dordrecht, 2000.

[33] H. W. Engl, A. Hofinger, and S. Kindermann. Convergence rates in the Prokhorov metric for assessing uncertainty in ill-posed problems. *Inverse Problems*, 21(1):399–412, 2005.

[34] M. A. T. Figueiredo, R. D. Nowak, and S. J. Wright. Gradient projection for sparse reconstruction: Applications to compressed sensing and other inverse problems. *IEEE Journal of Selected Topics in Signal Processing*, 1(4):586–597, 2007.

[35] M. Frazier, B. Jawerth, and G. Weiss. *Littlewood-Paley theory and the study of function spaces.* Number 79 in Regional Conference Series in Mathematics. American Mathematical Society, 1991.

[36] J.-J. Fuchs. On sparse representations in arbitrary redundant bases. *IEEE Transactions on Information Theory*, 50(6):1341–1344, 2004.

[37] J.-J. Fuchs. Recovery of exact sparse representations in the presence of bounded noise. *IEEE Transactions on Information Theory*, 51(10):3601–3608, 2005.

[38] J. Gire, L. Denis, C. Fournier, E. Thiébaut, F. Soulez, and C. Ducottet. Digital holography of particles: benefits of the 'inverse problem' approach. *Measurement Science and Technology*, 19(7):074005 (13pp), 2008.

[39] J. W. Goodman. *Introduction to Fourier optics.* Roberts & Co, Englewood, CO, 2005.

[40] M. Grasmair. Well-posedness and convergence rates for sparse regularization with sublinear l^q penalty term. *Inverse Problems and Imaging*, 3(3):383–387, 2009.

[41] M. Grasmair, M. Haltmeier, and O. Scherzer. Sparse regularization with ℓ^q penalty term. *Inverse Problems*, 24(5):055020 (13pp), 2008.

[42] M. Grasmair, M. Haltmeier, and O. Scherzer. Necessary and sufficient conditions for linear convergence of ℓ^1-regularization. Report 18, FSP S105, 2009.

[43] R. Gribonval and M. Nielsen. Beyond sparsity: Recovering structured representations by ℓ^1 minimization and greedy algorithms. *Advances in Computational Mathematics*, 28(1):23–41, 2008.

[44] R. Griesse and D. A. Lorenz. A semismooth Newton method for Tikhonov functionals with sparsity constraints. *Inverse Problems*, 24(3):035007 (19pp), 2008.

[45] T. Hein and K. S. Kazimierski. Accelerated Landweber iteration in Banach spaces. Technische Universität Chemnitz, Fakultät für Mathematik. Preprint 17, 2009.

[46] T. Hein and K. S. Kazimierski. Modified Landweber iteration in Banach spaces - convergence and convergence rates. Technische Universität Chemnitz, Fakultät für Mathematik. Preprint 14, 2009.

[47] K. D. Hinsch. Three-dimensional particle velocimetry. *Measurement Science and Technology*, 6(6):742–753, 1995.

[48] K. D. Hinsch and S. F. Hermann. Holographic particle image velocimetry. *Measurement Science and Technology*, 13(7):61–72, 2002.

[49] B. Hofmann, B. Kaltenbacher, C. Poeschl, and O. Scherzer. A convergence rates result for Tikhonov regularization in Banach spaces with non-smooth operators. *Inverse Problems*, 23(3):987–1010, 2007.

[50] J. A. Högbom. Aperture synthesis with a non-regular distribution of interferometer baselines. *Astronomy and Astrophysics Supplement Series*, 15:417–426, 1974.

[51] A. Jeffrey and D. Zwillinger. *Gradshteyn and Ryzhik's Table of Integrals, Series, and Products*. Academic Press, 7 edition edition, 2007.

[52] B. Jin, D. A. Lorenz, and S. Schiffler. Elastic-net regularization: Error estimates and active set methods. *Inverse Problems*, 25(11):115022 (26pp), 2009.

[53] B. Jin and J. Zou. Iterative schemes for Morozov's discrepancy principle in optimizations arising from inverse problems. Preprint, 2009.

[54] B. Kaltenbacher, F. Schöpfer, and T. Schuster. Iterative methods for nonlinear ill-posed problems in Banach spaces: convergence and applications to parameter identification problems. *Inverse Problems*, 25(6):065003 (19pp), 2009.

[55] K. S. Kazimierski. Minimization of the Tikhonov functional in Banach spaces smooth and convex of power type by steepest descent in the dual. *Computational Optimization and Applications*, 2009.

[56] K. C. Kiwiel. Proximal minimization methods with generalized Bregman functions. *SIAM Journal on Control and Optimization*, 35(4):1142–1168, 1997.

[57] E. Klann, M. Kuhn, D. A. Lorenz, P. Maass, and H. Thiele. Shrinkage versus deconvolution. *Inverse Problems*, 23(5):2231–2248, 2007.

[58] I. Krasikov. Uniform bounds for Bessel functions. *Journal of Applied Analysis*, 12(1):83–92, 2006.

[59] T. Kreis. *Handbook of holographic interferometry: optical and digital methods.* Wiley, 2005.

[60] L. J. Landau. Bessel functions: Monotonicity and bounds. *Journal of the London Mathematical Society*, 61(1):197–215, 2000.

[61] H. Lee, A. Battle, R. Raina, and A. Y. Ng. Efficient sparse coding algorithms. In B. Schölkopf, J. Platt, and T. Hoffman, editors, *Advances in Neural Information Processing Systems 19*, pages 801–808. MIT Press, Cambridge, MA, 2007.

[62] M. S. Lee and E. H. Kerns. LC/MS applications in drug development. *Mass Spectrometry Reviews*, 18(3–4):187–279, 1999.

[63] M. Liebling, T. Blu, and M. Unser. Fresnelets: new multiresolution wavelet bases for digital holography. *IEEE Transactions on Image Processing*, 12(1):29–43, 2003.

[64] D. A. Lorenz. Solving variational methods in image processing via projections - a common view on TV-denoising and wavelet shrinkage. *Zeitschrift für angewandte Mathematik und Mechanik*, 87(1):247–256, 2007.

[65] D. A. Lorenz. Convergence rates and source conditions for Tikhonov regularization with sparsity constraints. *Journal of Inverse and Ill-Posed Problems*, 16(5):463–478, 2008.

[66] D. A. Lorenz, S. Schiffler, and D. Trede. Beyond convergence rates: Exact inversion with Tikhonov regularization with sparsity constraints. arXiv:1001.3276, 2010.

[67] D. A. Lorenz and D. Trede. Optimal convergence rates for Tikhonov regularization in Besov scales. *Inverse Problems*, 24(5):055010 (14pp), 2008.

[68] D. A. Lorenz and D. Trede. Greedy deconvolution of point-like objects. In *Proceedings of the Second Workshop "Signal Processing with Adaptive Sparse Structured Representations"*, 2009.

[69] D. A. Lorenz and D. Trede. Optimal convergence rates for Tikhonov regularization in Besov scales. *Journal of Inverse and Ill-posed Problems*, 17(1):69–76, 2009.

[70] I. Loris. On the performance of algorithms for the minimization of ℓ^1-penalized functionals. *Inverse Problems*, 25(3):035008 (16pp), 2009.

[71] A. K. Louis. *Inverse und schlecht gestellte Probleme*. B.G. Teubner, 1989.

[72] S. Mallat. *A Wavelet Tour of Signal Processing*. Academic Press, 3 edition, 2009.

[73] S. Mallat, G. Davis, and Z. Zhang. Adaptive time-frequency decompositions. *SPIE Journal of Optical Engineering*, 33(7):2183–2191, July 1994.

[74] S. G. Mallat and Z. Zhang. Matching pursuits with time-frequency dictionaries. *IEEE Transactions on Signal Processing*, 41(12):3397–3415, 1993.

[75] K. Markides and A. Gräslund. Mass spectrometry (MS) and nuclear magnetic resonance (NMR) applied to biological macromolecules. Advanced information on the nobel prize in chemistry 2002. The Royal Swedish Academy of Sciences, 2002.

[76] R. E. Megginson. *An Introduction to Banach Space Theory.* Springer, 1998.

[77] H. Meng, G. Pan, Y. Pu, and S. Woodward. Holographic particle image velocimetry: from film to digital recording. *Measurement Science and Technology*, 15(4):673–685, 2004.

[78] F. Natterer. Error bounds for Tikhonov regularization in Hilbert scales. *Applicable Analysis*, 18(1&2):29–37, 1984.

[79] D. Needell and J. A. Tropp. CoSaMP: Iterative signal recovery from incomplete and inaccurate samples. *Applied and Computational Analysis*, 26(3):301–321, 2009.

[80] D. Needell, J. A. Tropp, and R. Vershynin. Greedy signal recovery review. In *Asilomar Conference on Signals, Systems, and Computers*, 2008.

[81] D. Needell and R. Vershynin. Uniform uncertainty principle and signal recovery via regularized orthogonal matching pursuit. *Foundations of Computational Mathematics*, 9(3):317–334, 2009.

[82] D. Needell and R. Vershynin. Signal recovery from inaccurate and incomplete measurements via regularized orthogonal matching pursuit. *IEEE Journal of Selected Topics in Signal Processing*, 4(2):310–316, 2010.

[83] Y. Pati, R. Rezaiifar, and P. Krishnaprasad. Orthogonal matching pursuit: recursive function approximation with applications to wavelet decomposition. In *Proceedings of 27th Asilomar Conference on Signals, Systems and Computers*, volume 1, pages 40–44, 1993.

[84] R. Ramlau. Regularization properties of Tikhonov regularization with sparsity constraints. *Electronic Transactions on Numerical Analysis*, 30:54–74, 2008.

[85] R. Ramlau and E. Resmerita. Convergence rates for regularization with sparsity constraints. Technical Report 2009-09, Johann Radon Institute for Computational and Applied Mathematics (RICAM), 2009.

[86] E. Resmerita. Regularization of ill-posed problems in Banach spaces: convergence rates. *Inverse Problems*, 21(4):1303–1314, 2005.

[87] E. Resmerita and O. Scherzer. Error estimates for non-quadratic regularization and the relation to enhancement. *Inverse Problems*, 22(3):801–814, 2006.

[88] A. Rieder. *Keine Probleme mit inversen Problemen.* Vieweg+Teubner, 2003.

[89] R. T. Rockafellar and R. J.-B. Wets. *Variational Analysis.* Springer, 1998.

[90] L. I. Rudin, S. J. Osher, and E. Fatemi. Nonlinear total variation based noise removal algorithms. *Physica D*, 60:259–268, 1992.

[91] T. Runst and W. Sickel. *Sobolev Spaces of Fractional Order, Nemytskij Operators, and Nonlinear Partial Differential Equations.* de Gruyter Series in Nonlinear Analysis and Applications. Walter de Gruyter, 1996.

[92] S. Schiffler. *The elastic net: Stability for sparsity methods.* PhD thesis, University of Bremen, 2010.

[93] U. Schnars and W. Jüptner. Principles of direct holography for interferometry. In W. Jüptner and W. Osten, editors, *FRINGE 93: Proc. 2nd Int. Workshop on Automatic Processing of Fringe Patterns*, 1993.

[94] U. Schnars and W. Jüptner. *Digital Holography.* Springer, 2005.

[95] F. Schöpfer, A. K. Louis, and T. Schuster. Nonlinear iterative methods for linear ill-posed problems in Banach spaces. *Inverse Problems*, 22(1):311–329, 2006.

[96] I. M. Singer. *Bases in Banach spaces I.* Springer, 1970.

[97] F. Soulez, L. Denis, C. Fournier, E. Thiébaut, and C. Goepfert. Inverse problem approach for particle digital holography: accurate location based on local optimisation. *Journal of the Optical Society of America A*, 24(4):1164–1171, 2007.

[98] F. Soulez, L. Denis, É. Thiébaut, C. Fournier, and C. Goepfert. Inverse problem approach in particle digital holography: out-of-field particle detection made possible. *Journal of the Optical Society of America A*, 24(12):3708–3716, 2007.

[99] E. M. Stein and G. Weiss. *Introduction to Fourier Analysis on Euclidean Spaces.* Princeton University Press, 1971.

[100] H. Triebel. *Theory of Function Spaces II.* Monographs in Mathematics. Birkhäuser, 1992.

[101] J. A. Tropp. Greed is good: Algorithmic results for sparse approximation. *IEEE Transactions on Information Theory*, 50(10):2231–2242, 2004.

[102] J. A. Tropp. Just relax: Convex programming methods for identifying sparse signals in noise. *IEEE Transactions on Information Theory*, 52(3):1030–1051, 2006.

[103] J. A. Tropp. Corrigendum in "Just relax: Convex programming methods for identifying sparse signals in noise". *IEEE Transactions on Information Theory*, 55(2):917–918, 2009.

[104] G. A. Tyler and B. J. Thompson. Fraunhofer holography applied to particle size analysis a reassessment. *Journal of Modern Optics*, 23:685–700, 1976.

[105] C. S. Vikram. *Particle field holography.* Cambridge University Press, 1992.

[106] C. R. Vogel and R. Acar. Analysis of bounded variation penalty methods for ill-posed problems. *Inverse Problems*, 10(6):1217–1229, 1994.

[107] J. T. Watson and O. D. Sparkman. *Introduction to Mass Spectrometry.* Wiley & Sons, fourth edition, 2007.

[108] Z. B. Xu and G. F. Roach. Characteristic inequalities of uniformly convex and uniformly smooth Banach spaces. *Journal of Mathematical Analysis and Applications*, 157(1):189–210, 1991.

[109] C. A. Zarzer. On Tikhonov regularization with non-convex sparsity constraints. *Inverse Problems*, 25(2):025006 (13pp), 2009.

[110] B. Zeff, D. Lanterman, R. McAllster, R. Roy, E. Kostelich, and D. Lathrop. Measuring intense rotation and dissipation in turbulent flows. *Nature*, 421(6919):146–149, 2003.

[111] H. Zou and T. Hastie. Regularization and variable selection via the elastic net. *Journal of the Royal Statistical Society, Series B*, 67(2):301–320, 2005.